Communications
in Computer and Information Science 489

T0210925

Heyan Huang   Ting Liu   Hua-Ping Zhang
Jie Tang (Eds.)

Social Media Processing

Third National Conference, SMP 2014
Beijing, China, November 1-2, 2014
Proceedings

 Springer

Volume Editors

Heyan Huang
Beijing Institute of Technology
School of Computer Science and Technology
Beijing, China
E-mail: hhy63@bit.edu.cn

Ting Liu
Harbin Institute of Technology
Department of Computer Science and Technology
Harbin, China
E-mail: tliu@ir.hit.edu.cn

Hua-Ping Zhang
Beijing Institute of Technology
School of Computer Science
Beijing, China
E-mail: kevinzhang@bit.edu.cn

Jie Tang
Tsinghua University
Department of Computer Science
Beijing, China
jietang@tsinghua.edu.cn

ISSN 1865-0929   e-ISSN 1865-0937
ISBN 978-3-662-45557-9   e-ISBN 978-3-662-45558-6
DOI 10.1007/978-3-662-45558-6
Springer Heidelberg New York Dordrecht London

Library of Congress Control Number: 2014953938

Typesetting: Camera-ready by author, data conversion by Scientific Publishing Services, Chennai, India

Printed on acid-free paper

Springer is part of Springer Science+Business Media (www.springer.com)

Preface

We are living in an increasingly networked world. People, information, and other entities are connected via the World Wide Web, e-mail networks, instant messaging networks, mobile communication networks, online social networks, etc. These generate massive amounts of social data, which present great opportunities in understanding the science of user behavioral patterns and the structure of networks formed by people interactions. The Third National Conference on Social Media Processing (SMP) was held in Beijing, China, in 2014 for the purpose of promoting original research in mining social media and applications, bringing together experts from related fields such as natural language processing, data mining, and information retrieval, and providing a leading forum in which to exchange research ideas and results in emergent social media processing problems.

The conference received 101 submissions, of which 49 were English submissions. All papers were peer reviewed by at least three members of the Program Committee (PC) composed of international experts in natural language processing, data mining, and information retrieval. The PC together with the PC co-chairs worked very hard to select papers through a rigorous review process and via extensive discussion. The competition was very strong; only 14 papers were accepted as full papers and nine as short papers. The conference also featured invited speeches from outstanding researchers in social media processing and related areas (the list may be incomplete): Yan Jia (National University of Defense Technology), Hang Li (Huawei Inc.), Dehuan Liu (Peking University), Huan Liu (Arizona State University), HuiXiong (Rutgers, the State University of New Jersey), and Qiang Ye (Harbin University).

Without the support of several funding agencies and industrial partners, the successful organization of SMP 2014 would not have been possible. Sponsorship was provided by the following companies, among others: Tencent, TRS, HYLANDA, SinoVoice, iFLYTEK, 360, WEIZOOM, DATATANG. We would also like to express our gratitude to the Steering Committee of the special group of Social Media Processing of the Chinese Information Processing Society for all their advice and the Organizing Committee for their dedicated efforts. Last but not least, we sincerely thank all the authors, presenters, and attendees who jointly contributed to the success of SMP 2014.

November 2014

Heyan Huang
Ting Liu
Hua-Ping Zhang
Jie Tang

Organization

Steering Committee Chair

Li Sheng — Harbin Institute of Technology

Steering Committee Co-chair

Li Yuming — Beijing Language and Culture University

Steering Committee

Bai Shuo	Shanghai Stock Exchange
Huang Heyan	Beijing Institute of Technology
Li Xiaoming	Peking University
Ma Shaoping	Tsinghua University
Meng Xiaofeng	Renmin University of China
Nie Jianyun	University of Montreal, Canada
Shi Shuicai	TRS
Sun Maosong	Tsinghua University
Wang Feiyue	Institute of Automation, Chinese Academy of Sciences
Zhou Ming	Microsoft Research Asia

President of the Assembly

Huang Heyan	Beijing Institute of Technology
Liu Ting	Harbin Institute of Technology

Vice-President of the General Assembly

Cheng Xueqi	Institute of Computing Technology, Chinese Academy of Sciences
Huang Xuanjing	Fudan University
Lin Hongfei	Dalian University of Technology
Zeng Dajun	Institute of Automation, Chinese Academy of Sciences

Program Committee Chair

Zhang Huaping	Beijing Institute of Technology
Tang Jie	Tsinghua University

Program Committee

Chen Yu	Defense Technology Information Center
Chen Zhumin	Shandong University
Ding Guodong	Institute of Computing Technology, Chinese Academy of Sciences
Dong Yi	Qihoo 360
Dou Yongxiang	Xd Management College of Information Management
Du Yuejin	National Institute of network Information Security Technology
Feng Chong	Beijing Institute of Technology
Feng Shizheng	Renmin University
Fu Guohong	Heilongjiang University
Gao Yue	National University of Singapore
Gao Ziguang	Tencent Microblogging
Gong Jibing	Yanshan University
Han Yi	Peking University
Han Qilong	Harbin Engineering University
Hao Xi Long	Massive Company
Hong Yu	Soochow University
Huang Jing	Renren Social Network
Ji Dong Hung	Wuhan University
Ji Zhong	Tianjin University
Jiang Wei	Beijing University of Technology
Jiang Sheng-yi	Guangdong University of Foreign Studies
Li Bing	University of International Business and Economics
Li Aiping	National University of Defense Technology
Li Guoliang	Tsinghua University
Li Juanzi	Tsinghua University
Lin Chunyu	TRS Corp.
Liu Kang	Institute of Automation, Chinese Academy of Sciences
Liu Yang	Shandong University
Liu Dexi	Jiangxi University of Finance and Economics
Liu Lizhen	Capital Normal University
Liu Shenghua	Institute of Computing Technology, Chinese Academy of Sciences
Liu Yiqun	Tsinghua University
Liu Zhiyuan	Tsinghua University
Ma Jun	Shandong University

Mao Wenji	Institute of Automation, Chinese Academy of Sciences
Mo Tong	School of Software and Microelectronics, Peking University
Peng Tao	College of Computer Science and Technology, Jilin University
Qi Haoliang	Heilongjiang Institute of Engineering
Qi Hongwei	Datatang
Qin Bing	Harbin Institute of Technology
Ruan Tong	East China University of Science and Technology
Sha Ying	Institute of Information Engineering, Chinese Academy of Sciences
Shen Hao	Communication University of Chin
Shen Yang	Wuhan University
Shen Huawei	Institute of Computing Technology, Chinese Academy of Sciences
Shi Hanxiao	Zhejiang Gongshang University
Song Wei	Capital Normal University
Sun Guanglu	Harbin University of Science and Technology
Turgen	Xinjiang University
Wang Bin	Institute of Computing Technology, Chinese Academy of Sciences
Wang Bo	Harbin Institute of Technology
Wang Lei	Institute of Automation, Chinese Academy of Sciences
Wang Ting	University of Defense Science and Technology
Wang Ying	College of Computer Science and Technology, Jilin University
Wang Zhen	Micro Focus Company
Wang Bailing	Harbin Institute of Technology
Wang Haixun	Microsoft Research Asia
Wang Jianxiong	Guangzhong University
Wang Mingwen	Jiangxi Normal University
Wang Shuaiqiang	Shandong University of Finance and Economics
Wang Zhenyu	South China Institute of Technology
Wu Hua	Baidu Inc.
Wu Dayong	Institute of Computing Technology, Chinese Academy of Sciences
Xia Yunqing	Research Institute of Information Technology, Tsinghua University
Xiong Jinhua	Institute of Computing Technology, Chinese Academy of Sciences
Xu Ruifeng	Harbin Institute of Technology

Xu Zhiming	HIT
Yang Xu	SAS Corp. USA
Yang Er Hong	Beijing Language and Culture University
Yang Hongwu	Institute of Automation, Chinese Academy of Sciences
Yang Zhihao	Dalian Institute of Technology
Ye Qiang	School of Management Science, HIT
Yin Jianmin	Weifang Huaguang
Yin Lan	Guizhou Normal University
Yuan Xiaoru	Peking University
Zhan Jian	Lanzhou University
Zhan Weidong	Peking University
Zhang Chuang	Beijing University of Posts and Telecommunications
Zhang Ming	Peking University
Zhang Peng	Institute of Information Engineering, Chinese Academy of Sciences
Zhang Qi	Fudan University
Zhang Shu	Fujitsu Limited
Zhang Yu	Harbin Institute of Technology
Zhang Guoqing	Institute of Computing Technology, Chinese Academy of Sciences
Zhang Huaping	Beijing Institute of Technology
Zhang Ziqiong	Harbin Institute of Technology
Zhang Chengzhi	Nanjing University of Science and Technology
Zhao Jun	Institute of Automation, Chinese Academy of Sciences
Zhao Shiqi	Baidu Inc.
Zhao Yanyan	Harbin Institute of Technology
Zheng Xiaolong	Institute of Automation, Chinese Academy of Sciences
Zhou Dong	Hunan University of Science and Technology
Zhou Xu	Chinese Academy of Sciences

Organizing Committee

Jian Ping	Beijing Institute of Technology
Shi Shumin	Beijing Institute of Technology
Shang Jianyun	Beijing Institute of Technology
Xin Xin	Beijing Institute of Technology
Zhao Yanping	Beijing Institute of Technology
Shi Xuewen	Beijing Institute of Technology
Sun Mengshu	Beijing Institute of Technology
Zhao Lianwei	Beijing Institute of Technology
Zhu Qian	Beijing Institute of Technology

Natural Language Processing for Social Media

Lin Hongfei   Dalian University of Technology
Zhan Weidong   Peking University

Social Network Analysis and Complex Systems

Cheng Xueqi   Institute of Computing Technology, Chinese
                               Academy of Sciences
Zeng Dajun   Institute of Automation, Chinese Academy
                               of Sciences

Social Media Processing and Social Sciences

Shen Hao   Communication University of China
Luo Jiade   Tsinghua University
Feng Shizheng   Renmin University of China

Social Media Mining, Forecasting and Recommendation

Huang Xuanjing   Fudan University

Social Media Analysis and Visualization

Lv Ke   China University Academy of Sciences

Social Media Public Opinion Analysis and Precise Marketing

Xiao Weidong   National University of Defense Technology

Social Media Security, Privacy Protection and Computing Support Platform

Zhang Yunquan   Institute of Computing Technology, Chinese
                               Academy of Sciences
Xu Jianliang   Hong Kong Baptist University
Chen Wenguang   Tsinghua University

Post Chair

Li Guoliang   Tsinghua University
Wang Bailing   Harbin Institute of Technology Weihai Campus

Financial Chair

Shi Shumin   Beijing Institute of Technology
Xiao Qianhui   Chinese Information Processing Society

Publications Chair

Zhao Yanping   Beijing Institute of Technology

Panel Chair

Lin Hongfei   Dalian University of Technology

Student Sponsorship Chair

Han Qilong   Harbin Engineering University
Jiang Shengyi   Guangdong University of Foreign Studies

Demo Chair

QI Haoliang   Heilongjiang Institute of Engineering
Xu Ruifeng   HIT Shenzhen Graduate School

Sponsorship Chair

Feng Chong   Beijing Institute of Technology
Wang Bin   Institute of Computing Technology, Chinese
                                    Academy of Science

Publicity Chair

Shen Hao   Communication University of China

Sponsors

The Third National Conference of Social Media Processing (SMP 2014) is committed to building a network of social media processing researchers, which is sponsored by the following organizations.

Table of Contents

Inferring Correspondences from Multiple Sources for Microblog User Tags

Cunchao Tu, Zhiyuan Liu, and Maosong Sun

Department of Computer Science and Technology
State Key Lab on Intelligent Technology and Systems
National Lab for Information Science and Technology
Tsinghua University, Beijing 100084, China
{tucunchao,lzy.thu}@gmail.com, sms@tsinghua.edu.cn

Abstract. Some microblog services encourage users to annotate themselves with multiple tags, indicating their attributes and interests. User tags play an important role for personalized recommendation and information retrieval. In order to better understand the semantics of user tags, we propose Tag Correspondence Model (TCM) to identify complex correspondences of tags from the rich context of microblog users. In TCM, we divide the context of a microblog user into various sources (such as short messages, user profile, and neighbors). With a collection of users with annotated tags, TCM can automatically learn the correspondences of user tags from the multiple sources. With the learned correspondences, we are able to interpret implicit semantics of tags. Moreover, for the users who have not annotated any tags, TCM can suggest tags according to users' context information. Extensive experiments on a real-world dataset demonstrate that our method can efficiently identify correspondences of tags, which may eventually represent semantic meanings of tags.

Keywords: User Tag Suggestion, Tag Correspondence Model, Probabilistic Graphical Model.

1 Introduction

As microblogs grow in popularity, Microblog users generate rich contents everyday, which include short messages and comments. Meanwhile, microblog users build a complex social network with following or forwarding behaviors. Both user generated content and social networks constitute the context information of a microblog user. In order to well understand the interests of users, some microblog services encourage users to label tags to themselves. Tags provide a powerful scheme to represent attributes or interests of microblog users, and may eventually facilitate personalized recommendation and information retrieval.

In order to profoundly understand user tags, it is intuitive to represent implicit semantics of user tags using correspondences identified from the rich context of microblog users. Here each **correspondence** is referred to a unique element in

H.-Y. Huang et al. (Eds.): SMP 2014, CCIS 489, pp. 1–12, 2014.

the context which is semantically correlated with the tag. For example, for the tag "mobile_internet" of Kai-Fu, we may identify the word "mobile" in his self description as a correspondence.

In general, the context information of microblog users origins from multiple **sources**. Each source has its own correspondence candidates. The sources can be categorized into two major types: **user-oriented ones** and **neighbor-oriented ones**.

To find precise correspondences of tags from these sources, two facts make it extremely challenging. (1) The context information is complex and noisy. For example, each user may generate many short messages with diverse topics and in informal styles, which makes it difficult to identify appropriate correspondences of tags. (2) The context information is from multiple and heterogenous sources, and each source has its own characteristics. It is non-trivial to jointly model multiple sources.

To address the challenges, we propose a probabilistic generative model, Tag Correspondence Model (TCM), to infer correspondences of user tags from multiple sources. Meanwhile, TCM can suggest tags for those users who have not annotated any tags according to their context information. For experiments, we build a real-world dataset and take user tag suggestion as our quantitative evaluation task. Experiment results show that TCM outperforms the state-of-the-art methods for microblog user tag suggestion, which indicates that TCM can efficiently identify correspondences of tags from the rich context information of users.

2   Related Work

There has been broad spectrum of studies on general social tag modeling and personalized social tag suggestion. Many studies have been done to suggest tags for products such as books, movies and restaurants[8,15,7,17,10]. These studies mostly focus on the tagging behaviors of a user on online items such as Web pages, images and videos.

As a personalized recommendation task, some successful techniques in recommender systems are introduced to address the task of social tag suggestion, e.g., user/item based collaborative filtering [14], matrix and tensor decomposition [18]. Some graph-based methods are also explored for social tag suggestion [8]. In these methods, a tripartite user-item-tag graph is built based on the history of user tagging behaviors, and random walks are performed over the graph to rank tags. We categorize these methods into the **collaboration-based approach**.

The above mentioned studies on social tag suggestion are all based on the history of tagging behaviors. There are also many researches focusing on recommending tags based on meta-data of items, which are usually categorized into the **content-based approach**. For example, some researchers consider each social tag as a classification category, and thereby address social tag suggestion as a task of multi-label classification [9]. In these methods, the semantic relations between features and tags are implicitly hidden behind the parameters of classifiers, and thus are usually not human interpretable.

Inspired by the popularity of latent topic models such as Latent Dirichlet Allocation (LDA) [2], various graphical methods are proposed to model the semantic relations of users, items and tags for social tag suggestion. An intuitive idea is to consider both tags and words as being generated from the same set of latent topics. By representing both tags and descriptions as the distributions of latent topics, it suggests tags according to the likelihood given the meta-data of items [16,11]. As an extension, [3] propose a joint latent topic model of users, words and tags. Furthermore, an LDA-based topic model, Content Relevance Model (CRM) [7], is proposed to find the content-related tags for suggestion, whose experiments show the outperformance compared to both classification-based methods and Corr-LDA [1], a typical topic model for modeling both contents and annotations.

Despite the importance of modeling microblog user tags, there has been little work focusing on this. Unlike other social tagging systems, in microblog user tagging systems each user can only annotate tags to itself. Hence, we are not able to adopt the collaboration-based approach. Since we want to interpret semantic meanings of user tags, the classification-based methods are not competent neither. Considering the powerful representation ability of graphical models, in this paper we propose Tag Correspondence Model (TCM). Although some graphical models have been proposed for other social tagging systems as mentioned above, most of them are designed for modeling semantic relations between tags and some *limited* and *specific* factors, such as users or words, and thus are not capable of joint modeling of rich context information. On the contrary, TCM can identify *complex* and *heterogeneous* correspondences of user tags from multiple sources. In our experiments, we will show that it is by no means unnecessary to consider rich context for modeling microblog user tags.

3 Tag Correspondence Model

We give some formalized notations and definitions before introducing TCM. Suppose we have a collection of microblog users U. Each user $u \in U$ will generate rich text information such as self description and short messages, annotate itself with a set of tags a_u from a vocabulary T of size $|T|$, and also build friendship with a collection of neighbor users f_u.

3.1 The Model

We propose Tag Correspondence Model (TCM) to identify correspondences of each tag from multiple sources of users including but not limited to self descriptions, short messages, and neighbor users. We design TCM as a probabilistic generative model.

We show the graphical model of TCM in Fig. 1. In TCM, without loss of generality, we denote all sources of a user as a set S_u. Each source $s \in S_u$ is represented as a weighted vector $x_{u,s}$ over a vocabulary space V_s. All elements in these vocabularies are considered as correspondence candidates. Each correspondence r from the source s is represented as a multinomial distribution $\phi_{s,r}$

over all tags in the vocabulary T drawn from a symmetric Dirichlet prior β. The annotated tags of a microblog user u is generated by first drawing a user-specific mixture π_u from asymmetric Dirichlet priors η_u, which indicates the distribution of each source for the user. For each source s, a user-specific mixture $\theta_{u,s}$ over V_s correspondences is drawn from asymmetric Dirichlet priors $\alpha_{u,s}$, which indicate the prior importance of correspondences for the user. Suppose $x_{u,s}$ indicates the normalized importance scores of all correspondences in the source s for the user u. We denote the prior of each correspondence r as $\alpha_{u,s,r} = \alpha x_{u,s,r}$, where α is the base score which can be manually pre-defined as in LDA [6].

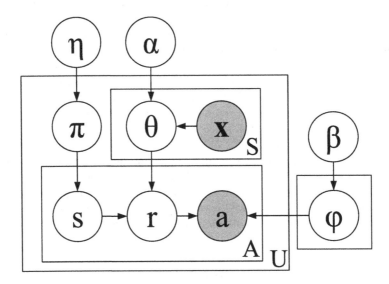

Fig. 1. Tag Correspondence Model

In TCM, the generative process of each tag t annotated by the user u is shown as follows: (1) picking a source s from π_u, (2) picking a correspondence r from $\theta_{d,s}$, and (3) picking a tag t from $\phi_{s,r}$. Hence, the tag t will be picked eventually in proportion to how much the user prefers the source s, how much the source s prefers the correspondence r, and how much the correspondence r prefers the tag t.

Note that one of these sources will be interpreted as a *global* source, which contains only one correspondence and presents on each user. When an annotated tag cannot find an appropriate correspondence from other sources, it will be considered as generated from the global correspondence.

In TCM, the annotated tags and the prior importance of correspondences in multiple sources are observed, which are thus shaded in Figure 1. We are required to find an efficient way to measure the joint likelihood of observed tags a and unobserved source and correspondence assignments, i.e. s and r, respectively.

The joint likelihood is formalized as follows,

$$\Pr(\boldsymbol{a}, \boldsymbol{s}, \boldsymbol{r} | \boldsymbol{x}, \alpha, \eta, \beta) = \Pr(\boldsymbol{a} | \boldsymbol{r}, \beta) \Pr(\boldsymbol{r}, \boldsymbol{s} | \boldsymbol{x}, \alpha, \eta). \tag{1}$$

By optimizing the joint likelihood, we will derive the updates for parameters of TCM including π, θ and ϕ. In this joint likelihood, the first item $\Pr(\boldsymbol{a}|\boldsymbol{r}, \beta)$ is similar to the word generation in LDA and thus we use the same derivation as in [6]. The second term can be decomposed as follows,

$$\Pr(\boldsymbol{r}, \boldsymbol{s} | \boldsymbol{x}, \alpha, \eta) = \Pr(\boldsymbol{r} | \boldsymbol{s}, \boldsymbol{x}, \alpha) \Pr(\boldsymbol{s} | \eta), \tag{2}$$

in which the two parts can be further formalized as

$$\Pr(\boldsymbol{s} | \eta) = \int_{\pi} \Pr(\boldsymbol{s} | \pi) \Pr(\pi | \eta) d\pi = \prod_{u \in U} \frac{\Delta(n_{u,:,:,\cdot} + \boldsymbol{\eta})}{\Delta(\boldsymbol{\eta})}, \tag{3}$$

$$\Pr(\boldsymbol{r} | \boldsymbol{s}, \boldsymbol{x}, \alpha) = \int_{\theta} \Pr(\boldsymbol{r} | \theta, \boldsymbol{s}) \Pr(\theta | \boldsymbol{x}, \alpha) d\theta = \prod_{u \in U} \prod_{s \in S} \frac{\Delta(n_{u,s,:,\cdot} + \boldsymbol{\alpha}_{u,s})}{\Delta(\boldsymbol{\alpha}_{u,s})}. \tag{4}$$

Here we denote the count $n_{u,j,k,t}$ as the number of occurrences of the source $j \in S_u$, the correspondence $k \in V_j$ as being assigned to the tag $t \in T$ of the user u. We further sum counts using "$\cdot$" and select a vector of counts using "$:$".

We observe that each correspondence is only allocated in one source, and thus there is no need to explicitly use the sources $\boldsymbol{s}$. We can use Gibbs Sampling to track the correspondence assignments $\boldsymbol{r}$. Following the derivations of LDA [6], the sampling update equation of assigning a new source and correspondence for a tag is formalized as follows,

$$\Pr(s_{u,i} = j, r_{u,i} = k | \boldsymbol{s}_{\neg u,i}, \boldsymbol{r}_{\neg u,i}, a_{u,i} = t, \alpha, \beta, \eta) \tag{5}$$

$$= \frac{n_{\cdot,j,k,t}^{(\neg u,i)} + \beta}{n_{\cdot,j,k,\cdot}^{(\neg u,i)} + |T|\beta} \cdot \frac{n_{u,j,\cdot,\cdot}^{(\neg u,i)} + (\boldsymbol{\alpha}_S)_j}{n_{u,\cdot,\cdot,\cdot}^{(\neg u,i)} + \sum_{j \in S}(\boldsymbol{\alpha}_S)_j} \cdot \frac{n_{u,j,k,\cdot}^{(\neg u,i)} + \alpha_{u,j,k}}{n_{u,j,\cdot,\cdot}^{(\neg u,i)} + \alpha_{u,j,\cdot}}$$

$$\propto \frac{n_{\cdot,j,k,t}^{(\neg u,i)} + \beta}{n_{\cdot,j,k,\cdot}^{(\neg u,i)} + |T|\beta} \cdot (n_{u,j,k,\cdot}^{(\neg u,i)} + \alpha x_{u,j,k}).$$

Here the sign $\neg u, i$ indicates that the count excludes the current assignment. For simplicity, we also define $(\boldsymbol{\alpha}_S)_j = \alpha_{u,j,\cdot}$, and thus the numerator in the second fraction cancels the denominator in the last fraction. Moreover, the denominator in the second fraction is constant for different source and correspondence assignment, and thus it is dropped in the last formula. We can observe that the update rule is quite similar to that of LDA.

For learning and inference, we can estimate the hidden parameters in TCM based on the collapsed sampling formula in Eq.(5). We can efficiently compute the counts n as the number of times that each tag has been assigned with each source and each correspondence. A sampler will iterate over the collection of users, reassign sources and correspondences, and update the counts. Finally, we can estimate the parameters of TCM using the source and correspondence assignments, in which we are mostly interested in

$$\pi_{u,s} = \frac{n_{u,s,\cdot,\cdot} + \eta}{n_{u,\cdot,\cdot,\cdot} + |S|\eta} \tag{6}$$

$$\theta_{u,s,r} = \frac{n_{u,s,r,\cdot} + \alpha x_{u,s,r}}{n_{u,s,\cdot,\cdot} + \alpha x_{u,s,\cdot}} \tag{7}$$

$$\phi_{s,r,t} = \frac{n_{\cdot,s,r,t} + \beta}{n_{\cdot,s,r,\cdot} + |T|\beta}. \tag{8}$$

3.2   Microblog User Tag Suggestion Using TCM

Given a user u with sources $s \in S$ and correspondences $r \in V_s$, the probability of selecting a tag t is formalized as

$$\Pr(t|u,\phi) = \sum_{s \in S} \sum_{r \in V_s} \Pr(t|r,\phi) \Pr(s,r|u) \Pr(s|u), \tag{9}$$

where $\Pr(s,r|u) = \Pr(r|u) = x_{u,s,r}$, and $\Pr(t|r,\phi) = \phi_{s,r,t}$. $\Pr(s|u)$ indicates the preference of each source s given the user u. Here we approximate $\Pr(s|u)$ using a global preference score of each source $\Pr(s)$, i.e. $\Pr(s|u) = \Pr(s)$. To compute $\Pr(s)$, we build a validation set to evaluate the suggestion performance with each source separately. By regarding the performance (e.g. F-Measure at $M = 10$ in this paper) as the confidence to the source, we assign $\Pr(s)$ as the normalized evaluation score of s. Then, we rank all candidate tags in descending order and select top ranked tags for suggestion.

4   Selection of Sources and Correspondences

We introduce in detail each source with its correspondences that will be used in TCM. We also define weighting measures for correspondences of each source, which will be used as prior knowledge x in Equation (5). In this paper, we consider two user-oriented sources: user messages(UM) and user descriptions(UD). We also consider two neighbor-oriented sources: neighbor tags(NT) and neighbor descriptions(ND). Inspired by term frequency and inverse document frequency (TF-IDF), we define some similar ways to measure the importance of each candidate in each source.

There are several methods incorporating network information into graphical models, such as Network Regularized Statistical Topic Model (NetSTM) [13] and Relational Topic Model (RTM) [4]. The basic idea of these methods is to smooth the topic distribution of a document with its neighbor documents. Although these methods provide an effective approach to intergrading both user-oriented and neighbor-oriented information, they suffer from two major issues. (1) These methods are not intuitively capable of modeling complex correspondences from multiple sources. (2) When modeling a document, the methods take its neighbor documents and their up-to-date topic distributions into consideration, which will be memory and computation consuming. Here we use a simple and effective way to model neighbor-oriented sources, whose effectiveness and efficiency will be demonstrated in our experiments.

5   Experiments and Analysis

We select Sina Weibo as our research platform. We randomly crawled 2 million users from Sina Weibo ranging from January 2012 to December 2012. From the raw data, we select $341,353$ users with each having complete profiles, short messages, social networks and more than 2 tags. In this dataset, the vocabulary size of tags is $4,126$. On average each user has 4.54 tags, 63.35 neighbors and 305.24 neighbor tags, and each user description has 6.93 words.

In TCM, we set $\beta = 0.1$ following the common practice in LDA [6] and set $\alpha = 10$ so as to leverage the prior knowledge of correspondence candidates.

In experiments, we use UM, UD, NT and ND to stand for the following four sources, user messages, user descriptions, neighbor tags and neighbor descriptions.

In order to intuitively demonstrate the efficiency and effectiveness of TCM, in Section 5.1 we perform empirical analysis of learning results, including characteristic tags and correspondences of TCM. Then in Section 5.2, we perform quantitative evaluation on TCM by taking user tag suggestion as the target application.

5.1   Empirical Analysis

Characteristic Tags of Sources. In order to better understand the four sources, in Table 1[1], we show the ratio of each source $\Pr(s)$ and Top-5 characteristic tags assigned to various sources. Here $\Pr(s)$ is computed by simply aggregating all source assignments for tags in U, i.e.

$$\Pr(s) = \frac{n_{\cdot,s,\cdot,\cdot} + \eta}{n_{\cdot,\cdot,\cdot,\cdot} + |S|\eta}. \tag{10}$$

We select representative tags of each source according to their characteristic scores in the source. Following the idea in [5], the characteristic score of a tag t in a source s is defined as

$$C(s,t) = \Pr(t|s) \times \Pr(s|t), \tag{11}$$

---

[1] To facilitate understanding, we explain some confusing tags as follows. "Fang Datong" is a Chinese pop-star. Chongqing, Shenzhen and Guangzhou are large cities in China. In the tag "Taobao Shopkeeper", Taobao is a popular c2c service. "Douban" is a book review service in China.

where

$$\Pr(t|s) = \frac{n_{\cdot,s,\cdot,t} + \beta}{n_{\cdot,s,\cdot,\cdot} + |T|\beta},$$

$$\Pr(s|t) = \frac{n_{\cdot,s,\cdot,t} + \beta}{n_{\cdot,\cdot,\cdot,t} + |S|\beta}.$$

Table 1. Proportion of each source and its characteristic tags. UM, UD, NT and ND stand for the following four sources, user messages, user descriptions, neighbor tags and neighbor descriptions.

Source	Pr(s)	Top 5 Characteristic Tags
UM	0.19	mobile internet, Fang Datong, Chongqing, Shenzhen, Guangzhou
UD	0.19	plane model, Taobao Shopkeeper, photographer, cosplay, e-business
NT	0.42	online shopping, novel, medium, reading, advertising
ND	0.20	Douban, lazy, novel, food, music

From the statistics in Table 1 we can see that neighbor-oriented sources are more important than user-oriented sources. What is more, the source of neighbor tags occupies the most important place in the four sources with a ratio of 0.42. The superiority of neighbor-oriented sources is not surprised. A user generates user-oriented content all by itself with much discretionary subjectivity, and thus may not necessarily fully reflect the corresponding user tags. Meanwhile, tags and descriptions of neighbors can be regarded, to some extent, as collaborative annotations to this user from their many friends, and thus may be more reasonable and less noisy.

Another observation from Table 1 is that, the characteristic tags of neighbor-oriented sources most reflect the interests of users, such as " online shopping", "reading", "food" and "music". On the contrary, most characteristic tags of user-oriented sources uncover the attributes of users, such as occupations, locations and identities. This indicates that, attribute tags may tend to find good correspondences from user-oriented sources, meanwhile interest tags from neighbor-oriented sources.

Note that, the setting of global source in TCM is important for modeling user tags. The global source collects the tags with no appropriate correspondences. Top 5 tags assigned to the global source are "music", "movie", "food", "80s" and "travel". These tags are usually general and popular, and have less correlations with the context information of users. If there is no global source, these tags will annoy the process of correspondence identification for other tags.

Characteristic Correspondences of Tags. The mission of TCM is to find appropriate correspondences for user tags. Here we pick some tags annotated

Table 2. Characteristic correspondences of Kai-Fu's tags

Tag	Top 5 Characteristic Correspondences
education	Internet (NT), eduction (UD), education (UM), politics (NT), study (NT)
technology	Android (NT), Internet (NT), product (ND), create (ND), communication (NT)
start-ups	start-ups (NT), venture capital (NT), e-business (NT), entrepreneur (NT), Internet (UD)
mobile internet	SNS (NT), mobile (UD), Internet (UM), mobile (UM), IT (NT)
e-business	B2C (NT), IT (NT), e-business (UM), e-business (NT), marketing (NT)

by Kai-Fu Lee as examples. In Table 2, we list characteristic correspondences of these tags. The characteristic score of a correspondence r with a tag t is computed as $C(r, t) = \Pr(t|r) \times \Pr(r|t)$. After each correspondence we provide the source in brackets. From these tags and their correspondences, it is convinced that TCM can identify appropriate correspondences from noisy and heterogeneous sources.

5.2   Evaluation on Microblog User Tag Suggestion

Evaluation Metrics and Baseline Methods. For the task of microblog user tag suggestion, we use precision, recall and F-Measure for evaluation. We also perform 5-fold cross validation for each method, and use the averaged precision, recall and F-Measure over all test instances for evaluation.

For microblog user tag suggestion, we select kNN [12], TagLDA [16], and NetSTM [13] as baseline methods for comparison. kNN is a typical classification algorithm based on closest training examples. TagLDA is a representative method of latent topic models, for which one can refer to [16] for detailed information. In this paper, we modify original NetSTM [13] by regarding tags as *explicit* topics, which can thus model the semantic relations between user-oriented contents with tags and take the neighbor tag distributions for smoothing. We set the number of topics $K = 200$ for TagLDA, the number of neighbors $k = 5$ for kNN, and the regularization factor $\lambda = 0.15$ for NetSTM, by which they obtains best performance.

Comparison Results. In Fig. 2 we show the precision-recall curves of different methods for microblog user tag suggestion. Here we use TCM-XX to indicate the method TCM using different sources, where UN indicates the combination of both user-oriented and neighbor-oriented sources. Each point of a precision-recall curve represents suggesting different number of tags from $M = 1$ to $M = 10$.

From Fig. 2 we observe that TCM significantly outperforms other baseline methods consistently except when it uses only short messages of users as the correspondence source. This indicates that the source of short messages in isolation is too noisy to suggest good user tags. We also find that TCM-UN achieves the best performance. When the suggestion number is $M = 10$, the F-Measure of TCM-UN is 0.184 while that of the best baseline method NetSTM is 0.142. This verifies the necessity of joint modeling of multiple sources for user tag suggestion.

In three baseline methods, kNN and Tag-LDA only consider the user-oriented source (i.e. self descriptions). The poor performance of kNN is not surprising because self descriptions are usually too short to computing appropriate user similarities. Although NetSTM models more sources with both neighbor tags and user descriptions, it goes behind Tag-LDA when suggesting more tags. This indicates that it is non-trivial to fuse multiple sources for user tag suggestion.

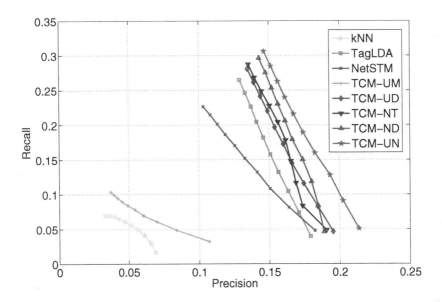

Fig. 2. Evaluation results of different methods

Note that, from Fig. 2 we find that the absolute evaluation scores of the best method TCM-UN are low compared with other social tagging systems [16,15]. This is mainly caused by the characteristics of microblog user tagging systems. On one side, since each user can only be annotated by itself, the annotated tags will be more arbitrary compared to other social tagging systems which are usually annotated collaboratively by thousands of users. On the other side, we perform evaluation by strictly matching suggested tags with user annotated tags. Hence, even a method can suggest reasonable tags for a user, which may usually have not been annotated by the specific user. Therefore, the evaluation scores can be used for comparing performance among methods, but are not applicable for judging the real performance of a method.

Case Study. In Table 3 we show top 5 tags suggested by TCM using various sources for the user Kai-Fu Lee we mentioned in Section 1. By taking the annotations of Kai-Fu as standard answers, we can see that most suggested tags are correct. What is more, although some suggested tags such as "Google", "marketing", "travel", "movie", and "reading" are not actually annotated by Kai-Fu, these tags are, to some extent, relevant to Kai-Fu according to his context. This also suggests that, even though the absolute evaluation scores of user tag suggestion are lower compared to some other research tasks, it does not indicate poor performance, but is caused by the strategy of complete matching with user annotations in evaluation.

Table 3. Tags suggested to Kai-Fu Lee from different sources

	Top 5 Suggested Tags
UM	mobile internet, start-ups, Internet, e-business, indoors-man
UD	innovation, freedom, Internet, Google, start-ups
NT	Internet, movie, start-ups, travel, e-business
ND	Internet, start-ups, e-business, marketing, mobile internet
UN	start-ups, e-business, Internet, mobile internet, reading

6 Conclusion and Future Work

In this paper, we formalize the task of modeling microblog user tags. We propose a probabilistic generative model, TCM, to identify correspondences as a semantic representation of user tags. In TCM we investigate user-oriented and neighbor-oriented sources for modeling, and carry out experiments on a real world dataset. The results show that TCM can effectively identify correspondences of user tags from rich context information. Moreover, as a solution to microblog user tag suggestion, TCM achieves the best performance compared to baseline methods.

We will explore the following directions as future work. (1) We will explore more rich sources to improve the performance of microblog user tag suggestion. (2) We will explore user factors for measuring $\Pr(s|u)$ when suggesting tags with TCM as shown in Section 3.2.

References

1. Blei, D., Jordan, M.: Modeling annotated data. In: Proceedings of SIGIR, pp. 127–134. ACM (2003)
2. Blei, D., Ng, A., Jordan, M.: Latent dirichlet allocation. JMLR 3, 993–1022 (2003)
3. Bundschus, M., Yu, S., Tresp, V., Rettinger, A., Dejori, M., Kriegel, H.: Hierarchical bayesian models for collaborative tagging systems. In: Proceedings of ICDM, pp. 728–733 (2009)
4. Chang, J., Blei, D.M.: Relational topic models for document networks. In: Proceedings of AISTATS, pp. 81–88 (2009)

5. Cohn, D., Chang, H.: Learning to probabilistically identify authoritative documents. In: Proceedings of ICML, pp. 167–174 (2000)
6. Griffiths, T., Steyvers, M.: Finding scientific topics. Proceedings of the National Academy of Sciences of the United States of America 101(suppl. 1), 5228–5235 (2004)
7. Iwata, T., Yamada, T., Ueda, N.: Modeling social annotation data with content relevance using a topic model. In: Proceedings of NIPS, pp. 835–843 (2009)
8. Jaschke, R., Marinho, L., Hotho, A., Schmidt-Thieme, L., Stumme, G.: Tag recommendations in social bookmarking systems. AI Communications 21(4), 231–247 (2008)
9. Katakis, I., Tsoumakas, G., Vlahavas, I.: Multilabel text classification for automated tag suggestion. In: ECML PKDD Discovery Challenge 2008, p. 75 (2008)
10. Liu, Z., Chen, X., Sun, M.: A simple word trigger method for social tag suggestion. In: Proceedings of the Conference on Empirical Methods in Natural Language Processing, pp. 1577–1588. Association for Computational Linguistics (2011)
11. Liu, Z., Tu, C., Sun, M.: Tag dispatch model with social network regularization for microblog user tag suggestion. In: 24th International Conference on Computational Linguistics, p. 755. Citeseer (2012)
12. Manning, C.D., Raghavan, P., Schütze, H.: Introduction to information retrieval, vol. 1. Cambridge University Press, Cambridge (2008)
13. Mei, Q., Cai, D., Zhang, D., Zhai, C.: Topic modeling with network regularization. In: Proceedings of WWW, pp. 101–110 (2008)
14. Peng, J., Zeng, D., Zhao, H., Wang, F.: Collaborative filtering in social tagging systems based on joint item-tag recommendations. In: Proceedings of CIKM, pp. 809–818. ACM (2010)
15. Rendle, S., Balby Marinho, L., Nanopoulos, A., Schmidt-Thieme, L.: Learning optimal ranking with tensor factorization for tag recommendation. In: Proceedings of KDD, pp. 727–736. ACM (2009)
16. Si, X., Sun, M.: Tag-LDA for scalable real-time tag recommendation. Journal of Computational Information Systems 6(1), 23–31 (2009)
17. Si, X., Liu, Z., Sun, M.: Modeling social annotations via latent reason identification (2010)
18. Symeonidis, P., Nanopoulos, A., Manolopoulos, Y.: Tag recommendations based on tensor dimensionality reduction. In: Proceedings of RecSys, pp. 43–50. ACM (2008)

Mining Intention-Related Products
on Online Q&A Community

Junwen Duan, Xiao Ding, and Ting Liu

Reseach Center for Social Computing and Information Retrieval,
Harbin Institute of Technology, Harbin, China
{jwduan,xding,tliu}@ir.hit.edu.cn

Abstract. User generated content on social media has attracted much attention from service/product providers, as it contains plenty of potential commercial opportunities. However, previous work mainly focuses on user Consumption Intention (CI) identification, and little effort has been spent to mine intention-related products. In this paper, we propose a novel approach to mine intention-related products on online Question & Answer (Q&A) community. Making use of the question-answer pairs as data source, we first automatically extract candidate products based on dependency parser. And then by means of the collocation extraction model, we identify the real intention-related products from the candidate set. The experimental results on our carefully constructed evaluation dataset show that our approach achieves better performance than two natural baseline methods. Our method is general enough for domain adaptation.

Keywords: Consumption Intention, Product Extraction, Q&A Community.

1 Introduction

People are used to conveying their needs and desires on social media platform. For example, "What toys are appropriate for children under three?" expresses the user's intention to buy toys. Previous works [1, 3, 5] mainly focus on the CI identification from user generated content, and little effort has been spent to mine intention-related products, possibly for the following reasons.

- Constructing such an intention-related product database requires a lot of domain-specific knowledge, especially domains like health care.
- It would be both time and labor consuming, and nearly impossible to construct a database containing all related products.

However, mining appropriate products to satisfy user's intention is important for product providers. If product providers could make immediate response to user intention, the sales volume would be greatly improved.

The development of online Q&A community provides new opportunities for solving this problem. Online Q&A community, as a product of the open and sharing Internet, is continuously gaining popularity. Famous online Q&A communities, such

H.-Y. Huang et al. (Eds.): SMP 2014, CCIS 489, pp. 13–24, 2014.
© Springer-Verlag Berlin Heidelberg 2014

as Yahoo! Answers and Baidu Knows, have accumulated millions of user-generated question-answer pairs [10], which can serve as a natural knowledge database. Consider the following scenario: a user posted a question on online Q&A community, e.g. *"what toys are appropriate for child under three"*, later someone else replied with a possible answer *"stuffed toys"*. Thus we can make a connection between the intention keyword *toy* and candidate product *"stuffed toys"*. After intention-related product identification process, we consider *"stuffed toys"* as intention-related product. Then whenever someone else presents similar intention, we could immediately recommend *"stuffed toys"* to him/her.

The problem is new and important, however, there are at least two main challenges.

- Social media text is notoriously noisy [6], how to extract products from the text accurately?
- Even with candidate products at hand, how can we identify the intention-related ones from the intention-unrelated ones without much human labor?

To address above challenges, in this paper, we propose a general framework to automatically mine intention-related products. Because CI identification is out of scope of this paper, we assume that the intention keyword set is available. We first make use of the online Q&A community question-answer pairs as data source and pick the pairs with intention keyword inside. We combine pattern-based method with dependency parser to automatically extract the candidate products from the answers. In order to identify the intention-related products, we further propose a novel approach based on collocation extraction model. We carefully construct an evaluation dataset and present our observations. The experimental results show that our approach achieves better performance than two baseline methods, i.e., *Co-occurrence* and *Jaccard Coefficient*.

The major contributions of our work are as follows.

- We make an attempt to mine knowledge from online Q&A community for commercial purpose, which has open up a new way to make use of online Q&A user-generated content.
- We first propose the task of intention-related products identification, and innovatively adopt collocation extraction model. Our approach achieves a remarkable performance without preprocessing.
- We carefully construct intention-related product dataset based on online Q&A community question-answering pairs.

This paper is organized as follows. Data and our observations are in Section 3. We then introduce our intention-related product mining approach in Section 4. And experiment setup and details are presented in Section 5. There, we also analyze the result. We then review the related approaches in Section 6. And finally in Section 7 we conclude our work and outlook possible future work.

2   Problem Statement

Intention Keyword. Intention keyword is a single word or a phrase that can indicate user's CI most, and a user post with CI must at least contain one intention keyword. According to [5], CI could be further classified into explicit and implicit ones, thus a CI is the desire that explicitly or implicitly expressed by consumer to buy something.

Example 1: my baby is calcium deficient, what can I do?

Example 1 shows a possible post with intention keyword *"calcium deficient"*, because it may give rise to the user's further action to buy calcium supplement products.

Intention-Related Products. Intention-related products are the products that consumers with such intention would like to buy. Here products that we extracted consist of the following two different levels.

- Full product name, which refers to a certain product e.g. JNJ baby soap (强生婴儿香皂);
- Product category, which refers to a category of products, e.g. milk powder (奶粉).

Our approach takes both levels into consideration. Because we notice that some consumers are quite sure what products could satisfy their intention, and we could recommend certain products to them. However, others may not have a clear product in mind, thus we could recommend a category of products.

In this paper, we only focus on intention-related products mining in the Baby & Child Care domain. If not specially stressed, all the running cases are in this domain. And we suggest that the domain picked should satisfy the following conditions:

- The intention in the domain must be general and urgent, thus they can arouse hot discussion;
- Do not require too much domain-specific knowledge, so it would be easier to evaluate.

3   Data and Observations

As the task is first proposed by us, there is no available dataset for experiment and evaluation. In this section, we first introduce how we construct our experiment and evaluation dataset. We then present our observations over the dataset.

3.1   Data Collection

We carefully pick three most famous Baby & Child Care websites in Chinese, namely Taobao Wenda[1], BabyTree[2] and Sina Baby & Child Care Q&A[3].We start our web

---

[1] Taobao Wenda: http://baobao.wenda.taobao.com
[2] BabyTree: http://www.babytree.com
[3] Sina Baby & Child Care Q&A: http://ask.baby.sina.com.cn/

spider to crawl the question-answer pairs from the three websites, and obtain more than 700 thousand pairs, which nearly cover all aspects of Baby & Child Care.

Intention keyword set. We first randomly choose 4000 questions from the question-answer dataset, and manually annotate the intention keywords inside them. After that, we have obtain1380 questions with intention keywords. There remain 245 after removing the duplicated keywords. We then randomly pick 30 intention keywords and they make up the intention keyword set. Table 1 lists nine intention keywords in our set.

Table 1. Intention keyword samples

便秘 (constipation)	磨牙 (molar)	消毒 (disinfect)
吃手 (eat hands)	冲奶 (mix milk powder)	缺钙 (calcium deficient)
学步 (learn to walk)	枕秃 (pillow baldness)	拉稀 (diarrhea)

Intention-related product standard set. For each keyword in the intention keyword set, we look it up in the question-answer pairs. After that we obtain question-answer pairs with intention keyword in the question. Via method that will be mentioned in 4.1, we obtain the candidate products for each intention keyword. We hire two annotators to annotate the candidate products as related and unrelated to the intention keyword, and the agreement between annotators is measured using Cohen's Kappa Coefficient [2]. We only keep the candidate products that both annotators judged as intention-related. Due to space limit, we don't present the annotation guidelines here. We obtain highest Kappa = 0.86 and lowest Kappa = 0.78, which is substantial. As a result, we construct an intention-related product standard.

3.2   Observations

Table 1 shows the statistics of our constructed dataset, from which we can find that intention-related products only account for less than 10% in all candidate products extracted.

Table 2. Statistics of constructed dataset

Intention keywords	30
Average candidate products per intention keyword	345.7
Average intention-related products per intention keyword	33.3

We review the candidate products and discover that products judged as intention-unrelated could be further divided into two categories.

- *Not a real product*, this is due to limitation of pattern-based method and poor performance of dependency parser. Text matches the pattern will be extracted as a product and precision of dependency parser fall sharply in noisy social media text.
- *Intention-unrelated*, although the product has been mentioned in the answer, however, it has nothing to do with the intention.

This accounts for why we further carry out the intention-related product identification process.

We further make statistics of distribution of intention keyword and intention-related products. Figure 1 presents the distributions of intention keyword on our dataset. We can find that a majority of intention keyword has a high frequency (occurred more than 500 times in questions); this is the foundation of our candidate product extraction. Figure 2 shows the distribution of intention-related products. According to Table 1, on average per intention keyword relates to 33.3 products, from Figure 2 we can further discover that most intention keywords have more than 20 related products, which ensures the diversity of products could be recommended.

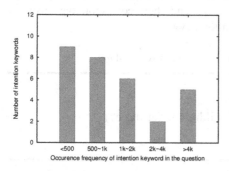

Fig. 1. Distribution of intention keywords on our dataset

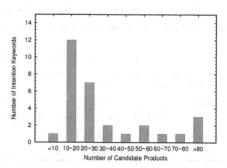

Fig. 2. Distribution of intention-related products on our dataset

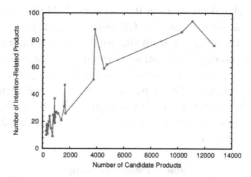

Fig. 3. Relation between occurrence frequency of an intention keyword and amount of its related products

Does more frequent occurrence of an intention keyword in questions leads to more related products? Our answer is "most times it does, but not always", which accords with our common knowledge. Figure 3 shows the relation between occurrence frequency of an intention keyword and amount of its related products. Frequent occurrence means more candidate products extracted, thus more related products identified. However, some intention keyword has a quite limited related product set, which means increase in occurrence frequency does not lead to more related products. However, with more candidate products at hand, we can cover more related products.

4   Approach

Our approach consists of three steps. First, we extract user CI. In this paper, we suppose user CI keyword set is available. Second, we identify candidate product names based on dependency parser. Third, we extract intention-related products. The details of each component are introduced as follows.

4.1   Candidate Product Extraction

Given an intention keyword k, we have to search from the question-answer pairs for questions that contain the intention keyword k and their corresponding answers. Table 2 shows an illustrative sample for intention keyword *toy*.

Table 3. An illustrative sample for intention keyword *toy*

Intention Keyword: *toy*
Question: 什么玩具适合三岁以下的儿童呢？
(What toys are appropriate for children under three?)
Answer: 我推荐毛绒玩具。
(I recommend stuffed toys.)

We adopt pattern-based method introduced in [5] to extract all possible products occurred in answers. Patterns are constructed based on our observations on how people recommend related products in the answers. Table 3 shows some patterns we use in our method. Our work is based on Chinese corpus, and we further apply semantic analysis to the answers apart from the part-of-speech information. We use the LTP[4] (Language Technology Platform) to carry out the language analysis. LTP [2] is a platform that has integrated Chinese word segmentation, part-of-speech tagger and semantic dependency parser. The dependency relations we make use of are VOB (verb-object), COO (coordinate) and SBV (subject-verb), by analyzing the features of each relation, we can easily obtain the products inside it. Table 4 demonstrates sample for each dependency relation.

Table 4. Pattern and Sample

Pattern	Sample
试试(try)	为什么不试试贝亲的奶瓶？
	Why not try milk bottle by Pigeon?
买(buy)	我买的是妈咪宝贝的纸尿裤。
	I bought diapers by Mamy Poko.
建议(suggest)	我推荐来自德国的SINA积木。
	I suggest puzzle blocks from German SINA.

---

[4] LTP: http://www.ltp-cloud.com

Table 5. Dependency Relation and Sample

Dependency Relation	Sample
VOB(verb-object)	Root 我 宝宝 用 贝亲 奶嘴 。 Root My baby use Pigeon pacifier .
COO(coordinate)	Root 毛绒 玩具 、 积木 和 芭比 娃娃 。 Root Stuffed toys 、 building blocks and Barbie doll .
SBV(subject-verb)	Root 妈咪 宝贝 的 纸尿裤 不错 ！ Root Mommy Poko 's disapers are good ！

4.2 Intention-Related Product Identification

Via the product extraction step, we have obtained a dataset containing the intention keywords and their corresponding candidate products. We view the intention-related product identification problem as a collocation evaluation problem. An intention keyword and a product form a collocation means that the intention keyword has once appeared in the question and the product have been extracted from its corresponding answer. Note that we may extract the same product from an answer for multi times, however, it would be only taken into account for once.

An intention keyword may collocate with many products, and a product as well may collocate with many intention keywords. A product with higher collocation probability to the intention keyword is more likely to be the intention-related product. Thus we can make use of the method in [7] to deal with intention-related product identification. Based on the extraction result, we can figure out the frequency that intention keyword k_i collocates with candidate product p_j, denoted as $freq(k_i, p_j)$, and then we estimate the probability that k_i collocates with p_j using Eq. (1), and the probability that p_j collocates with k_i using Eq. (2). The average collocation proba-bility of product p_j to intention keyword k_i is calculated using Eq. (3).

$$p(p_j \mid k_i) = \frac{freq(k_i, p_j)}{freq(k_i)} \tag{1}$$

$$p(k_i \mid p_j) = \frac{freq(k_i, p_j)}{freq(p_j)} \tag{2}$$

$$\overline{p}(k_i, p_j) = \frac{p(k_i \mid p_j) + p(p_j \mid k_i)}{2} \tag{3}$$

However, in the experiment we noticed that collocations with low frequency may achieve a high collocation probability under circumstance that product p_j is infrequent. In order to penalize the collocations with low frequency, we added a penalization factor, thus the intention-relatedness score is calculated by Eq. (4) .Here b is a constant parameter, according to central-limit theorem[4], sample set size larger than 30   is sufficiently large, thus, we set $b = 5.9$.

$$\overline{p}(k_i, p_j) = \frac{p(k_i \mid p_j) + p(p_j \mid k_i)}{2} \times e^{\frac{-b}{\log(freq(k_i, p_j)) + 1}} \tag{4}$$

5   Experiments

5.1   Baseline Methods

To evaluate the effectiveness of our proposed approach, we compare it with the following two baseline methods.

- Co-occurrence. To evaluate the intention-relatedness, the first method we may come up with is the co-occurrence-based one for the simple idea that the more a candidate product co-occurs with the intention keyword, the more likely it is the intention-related product. Here co-occurrence means the intention keyword appears in the question and candidate product in its corresponding answer.The co-occurrence score is calculated using Eq. (5)

$$SCORE_{cooccurence}(k_i, p_j) = freq(k_i, p_j) \tag{5}$$

- Jaccard Coefficient. Jaccard Coefficient is generally used to evaluate the similarity of two sets. However, we adopt it here to calculate the intention-relatedness. Based on the idea that the more similar the distribution of a candidate product to the distribution of an intention keyword, or the extreme case that the candidate product only co-occurs with the intention keyword, the more likely the candidate product is an intention-related product. The Jaccard Coefficient is calculated using Eq. (6).

$$SCORE_{Jaccard}(k_i, p_j) = \frac{freq(k_i, p_j)}{freq(k_i) + freq(p_j) - freq(k_i, p_j)} \tag{6}$$

5.2   Evaluation Metrics

We adopt two assessment metrics, namely *R-Precision* (Eq.7) and *WARP* (*Weighted Average R-Precision*) (Eq.8), to measure the performance. Before we get evolution started, we calculate and rank the scores of candidate products using our method and baseline method.

R-precision measures the precision in top *R* ranked candidate products, or in another word, how many candidate products are related to the intention keyword k_i in Top *R*. Because our data is highly skewed, according to previous observations, some of the intention keyword is much more frequent than others; as a result, they will have more candidate products and thus are more likely to have related products. Therefore, we use *WARP*, which takes frequency information into account, to calculate average *R-Precision*.

$$R\text{-Precision}(k_i) = \frac{\#\text{Related products to } k_i \text{ in top } R}{R} \qquad (7)$$

$$WARP(k_i) = \sum_i \frac{freq(k_i) \cdot R\text{-precision}(k_i)}{\sum_i freq(k_i)} \qquad (8)$$

5.3   Results and Analysis

Table 6 demonstrates TOP *R* precision of MWA and baseline methods on two randomly picked intention keyword. We can find that MWA generally outperforms the two baselines, and the superiority become even more obvious as *R* grows. What's more, unlike *Jaccard* and *Co-occurrence* based methods, they drop sharply as R increases, while precision of MWA keeps even. Figure 4 shows the *WARP* of our method and the baseline methods, from which we can find that our method is remarkable better than the baseline methods, more than 20% higher *WARP* than the best baseline method. Note that our result is achieved automatically without preprocessing the candidate product sets. With a maximum *WARP* of 63.3%, our method could meet the requirement of application.

Our method is better because we innovatively regard it as a collocation identification problem. We notice that the candidate products extracted contains a lot of noise, including unrelated products noisy text. Simply considering co-occurrence information could not get rid of the noise. Thus instead of focusing on the frequency, we take both the specificity a product to an intention keyword and the specificity an intention keyword to a product into consideration. What's more, we add a penalization factor to deal with infrequent cases.

Table 6. Performance of MWA and baseline method on intention keyword *calcium deficient* and *diarrhea*

Keyword	Method	P@5	P@10	P@15	P@20	P@25	P@30
缺钙	MWA	**0.80**	**0.80**	**0.73**	**0.70**	**0.68**	**0.63**
(calcium	Jaccard	0.60	0.70	0.53	0.45	0.40	0.33
deficiency)	Co-occurrence	0.60	0.60	0.47	0.40	0.40	0.33
拉稀	MWA	0.00	**0.30**	**0.27**	**0.25**	**0.24**	**0.30**
(diarrhea)	Jaccard	0.20	0.10	0.13	0.15	0.16	0.20
	Co-occurrence	0.00	0.10	0.13	0.20	0.20	0.20

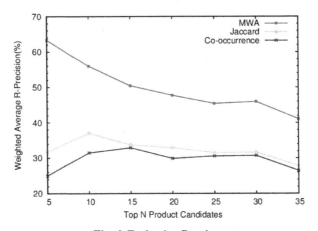

Fig. 4. Evaluation Result

We also notice that Jaccard-based method is not much better than occurrence-based one, because candidate products in TOP R have similar distribution so that it's difficult to tell them apart, or in another word, more similar distribution does not mean higher intention-relatedness.

6   Related Work

Our work is related to the following:

Text Mining on Social Media. Social media as a real-time data source, a great deal of text mining work have been done on it. [13] proposes to summarize twitter content by extracting key phrases. [9] starts an experimental study to extract name entity from noisy tweets. [8] focuses on sentiment analysis and opinion mining based on twitter corpus. [11] makes use of real-time nature of twitter, and use each user as a sensor to extract earthquake event and report earthquake .

Online Consumption Intention Identification. With the increasing popularity of online communities, CI identification has long attracted attentions from researchers. [3] first gives a formal definition of online commercial intention(OCI), and proposed a supervised method to predict whether submitting a search query or visiting a web-page will lead to commercial activity. Later work [1] studied the relationship between query terms and ad click behavior. Making use of query log and ad click data, they apply Bayes Thorem to quantify how much a term in a query contributes to underly-ing commercial intention. [5] first starts the task of CI detection on social media, and classify CI into explicit and implicit ones. They learned a classification model using word and part-of speech n-grams.

The most related work to ours is [12]; they try to mine trend-driven CI. However, we have different focus; they focus on trend-related products mining on microblog platform while we focus on intention-related products mining on online Q&A community. What's more, trend-driven CI has a natural evaluation metric which is product sales.

7   Conclusion and Future Work

We make our own attempt to mine intention-related products on online Q&A com-munity. Given a set of intention keywords, we automatically extract the candidate products from Q&A pairs. Our intention-related product identification method bor-rowed from collocation identification achieved a quite satisfying performance even on noisy data. The method describe in the paper could be further integrated into recom-mendation system. Our future work includes the following:

- Integrate with intention identification module, so that both intention identification and intention-related products recommendation could be done automatically;
- Add filter process after candidate product extraction, since many of the "products" extracted are not real products, by removing these "products" from the candidates, the precision of intention-related product identification would be greatly improved;
- Make use of the purchase data of intention-related products so that we can recom-mend more related products, even those not occurred in the candidate product set.

Acknowledgments. We thank the anonymous reviewers for their constructive comments, and gratefully acknowledge the support of the National Basic Research Program (973 Program) of China via Grant 2014CB340503, the National Natural Science Foundation of China (NSFC) via Grant 61133012 and 61202277.

References

1. Ashkan, A., Clarke, C.L.: Term-based commercial intent analysis. In: Proceedings of the 32nd International ACM SIGIR Conference on Research and Development in Information Retrieval, pp. 800–801. ACM (July 2009)
2. Che, W., Li, Z., Liu, T.: Ltp: A Chinese language technology platform. In: Proceedings of the 23rd International Conference on Computational Linguistics: Demonstrations, pp. 13–16. Association for Computational Linguistics (August 2010)

3. Dai, H.K., Zhao, L., Nie, Z., Wen, J.R., Wang, L., Li, Y.: Detecting online commercial intention (OCI). In: Proceedings of the 15th International Conference on World Wide Web, pp. 829–837. ACM (May 2006)
4. Grinstead, C.M., Snell, J.L.: Introduction to probability. American Mathematical Soc. (1998)
5. Hollerit, B., Kröll, M., Strohmaier, M.: Towards linking buyers and sellers: Detecting commercial Intent on twitter. In: Proceedings of the 22nd International Conference on World Wide Web Companion, pp. 629–632. International World Wide Web Conferences Steering Committee (May 2013)
6. Kaufmann, M., Kalita, J.: Syntactic normalization of twitter messages. In: International Conference on Natural Language Processing, Kharagpur, India (July 2010)
7. Liu, Z., Wang, H., Wu, H., Li, S.: Collocation extraction using monolingual word alignment method. In: Proceedings of the 2009 Conference on Empirical Methods in Natural Language Processing, vol. 2, pp. 487–495. Association for Computational Linguistics (August 2009)
8. Pak, A., Paroubek, P.: Twitter as a Corpus for Sentiment Analysis and Opinion Mining. In: LREC (May 2010)
9. Ritter, A., Clark, S., Etzioni, O.: Named entity recognition in tweets: An experimental study. In: Proceedings of the Conference on Empirical Methods in Natural Language Processing, pp. 1524–1534. Association for Computational Linguistics (July 2011)
10. Shah, C., Pomerantz, J.: Evaluating and predicting answer quality in community QA. In: Proceedings of the 33rd International ACM SIGIR Conference on Research and Development in Information Retrieval, pp. 411–418. ACM (July 2010)
11. Sakaki, T., Okazaki, M., Matsuo, Y.: Earthquake shakes Twitter users: Real-time event detection by social sensors. In: Proceedings of the 19th International Conference on World Wide Web, pp. 851–860. ACM (April 2010)
12. Wang, J., Zhao, W.X., Wei, H., Yan, H., Li, X.: Mining New Business Opportunities: Identifying Trend related Products by Leveraging Commercial Intents from Microblogs. In: EMNLP, pp. 1337–1347 (2013)
13. Zhao, W.X., Jiang, J., He, J., Song, Y., Achananuparp, P., Lim, E.P., Li, X.: Topical keyphrase extraction from twitter. In: Proceedings of the 49th Annual Meeting of the Association for Computational Linguistics: Human Language Technologies, vol. 1, pp. 379–388. Association for Computational Linguistics (June 2011)

Tag Expansion Using Friendship Information: Services for Picking-a-crowd for Crowdsourcing

Bin Liang, Yiqun Liu, Min Zhang, Shaoping Ma, Liyun Ru, and Kuo Zhang

State Key Laboratory of Intelligent Technology and Systems
Tsinghua National Laboratory for Information Science and Technology
Department of Computer Science and Technology
Tsinghua University, Beijing, China, 100084
mgigabyte@gmail.com, {yiqunliu,z-m,msp}@tsinghua.edu.cn,
{ruliyun,zhangkuo}@sogou-inc.com

Abstract. To address self-tagging concerns, some social networks' websites, such as LinkedIn and Sina Weibo, allow users to tag themselves as part of their profiles; however, due to privacy or other unknown reasons, most of the users take just a few tags. Self-tag sparsity refers to the problem of low recall obtained when searching for people on systems based on user profiles. In this paper, we use not only users' self-tags but also their friend relationships (which are often not hidden) to expand the tag list and measure the effectiveness of different types of friendship links and their self-tags. Experimental results show that friendship information (friendship links and profiles) can effectively improve the performance of tag expansion, especially for common users who have limited followers.

Keywords: Tag expansion, Self-tag mining, Energy function, Machine learning.

1 Introduction

With the boom in social networks, microblog services such as Twitter, Sina Weibo and LinkedIn have grown rapidly. On Mar. 21st, 2012, its sixth birthday, Twitter announced that it had 140 million users and 340 million tweets per day. On Feb. 20th, 2013, Sina Microblog announced that it had 500 million users and 46 million active users per day. With the development of social networks, many applications are based on the profiles and social relationships of these millions of users (G. A. Gupta 2013, Liang Bin 2014). In practice, users are willing to create profiles on these online social networks; the profiles consist of attributes such as location, hobbies, and sex. Additionally, users love to follow other users depending on their interests. This phenomenon is called *homophily*, which is a tendency that says "interpersonal similarity breeds connection" (M. McPherson 2011). Weng has reported that users who follow each other reciprocally usually share topical interests (Weng 2010).

It is, however, a heavy burden to require every user to create a complete profile. Although many social network sites allow users to tag themselves with a few

H.-Y. Huang et al. (Eds.): SMP 2014, CCIS 489, pp. 25–43, 2014.

keywords to reduce the burden, many users still take only a few tags for themselves, due to privacy or other reasons.

As a result, we have two types of data: user following data and user profile data. Because the former is sparse and the latter is rich, it guides us to apply user following data to predict user profiles. Actually, the ability to automatically predict user attributes could be useful for a variety of social networking applications such as friend and content recommendations (G. A. Gupta 2013) and our people search system[1], a research project on picking-a-crowd from social networks for crowdsourcing. Here, we would like to briefly introduce crowdsourcing. Current approaches to crowdsourcing are viewed as a type of pull methodology, where tasks are split and published on platforms where online workers can pick their preferred tasks. In fact, this type of approach has many advantages, such as simplicity and equality; however, it does not guarantee the assigning of tasks to suitable workers. We provide a service for crowdsourcing based on a push methodology that carefully selects workers to perform given tasks according to their profiles extracted from social networks. As a result, the self-tag sparsity problem makes it hard for our system to find enough candidate workers to complete a crowdsourcing task.

Workers on crowdsourcing platforms are neither celebrities nor well-known users but common users; therefore, expanding the tags of these users is a major challenge and must rely on automatic algorithms. It occurred to us that while common users may post a few microblogs and take a few tags, they usually maintain a good social network. Therefore, we mainly focus on expanding the tags of common users using their friendship information and their self-tags.

To summarize our motivation, our work focuses on expanding tags for users, especially common users, in social networks. Many users take only a few tags for themselves; this makes some of them unsearchable, and the quantity and quality of candidate workers in crowdsourcing systems is not up to the mark.

Before we introduce our model, we would like to briefly explain the differences between social tagging, people-tagging and self-tagging. Social tagging is a way for users to freely choose keywords to describe Internet content resources (Delicious and Flickr provide the service). People-tagging is a form of social bookmarking that enables people to organize their contacts into groups, annotate them with terms supporting future recall, and search for people by topic area (Bernstein 2009; Farrell 2007). Self-tagging is a way for users to tag themselves, for example, LinkedIn and Sina Weibo are services that allow users to only tag themselves but do not allow them to tag other users. Despite their differences, the above three concepts still have many similarities. Social tagging, People tagging and Self-tagging all aim at getting better descriptions of an object to make it easy to search and share. Muller's work shows that self-tags usually reflect the hobbies, knowledge-domain, location and social role of a user, which is the same as social tags (Muller 2006).

We now introduce the baseline and our model. In this paper, we employ association rules mining (Heymann 2008), a tag recommendation approach based on joint probability (Rae 2010) and the random walk algorithm (Li 2009) as our baseline algorithms. We do a survey to explore sources of tag expansion and discover that users'

---

[1] Our online people search system, http://xunren.thuir.org/

self-tags have the shortest KL divergence to the tags of their bidirectionally linked friends and the largest KL divergence to the keywords of microblogs that users post. We only choose as features the tag frequency of users' social relationships, the conditional probability of friends' tags given users' self-tags and the prior probability of tags and adopt an energy-based function to create our model and use negative log-likelihood loss as our loss function to train our model; this approach leads us to discover that *Precision* and *Recall* of tag expansion have improved significantly. Moreover, we also discuss the differences between these improvements for common users and celebrities.

We share the data related to our paper on a web page for researchers[2].
Finally, we sum up our contributions.

- We take users' self-tags and 3 types of users' friendship information into consideration to expand tags, as detailed in Section 4, and we show the effectiveness of different kinds of friendship information by experiments, as detailed in Section 5.
- We are the first to use different ranges of followers to delineate the model's performance on common users vs. celebrities, as detailed in Section 5, and we show that our algorithm is effective on common users who have fewer followers.
- We discuss the power of friendship information exacting on the performance of tag expanding.

The remainder of this paper is organized as follows: we introduce background information and related work in Section 2 and define the problem in Section 3. Then, Section 4 describes in depth our work including our survey, the related baseline chosen and our algorithms. Section 5 illustrates experiments showing the performance of each algorithm on sets having different ranges of users' followers. Section 6 includes some discussion on three questions on our algorithm and baseline. Finally, the summary is presented in Section 7.

2   Related Work

Many research efforts focus on tag expansion (tag recommendation or tag suggestion); however, these works mainly serve applications that make objects easy to be searched and shared, such as expanding tags of photos (Kucuktunc, 2008; Garg, 2008; Li, 2009), MP3s (Eck, 2007) and Blog posts (Sood, 2007). However, as far as we know, few works focus on expanding tags for linked people.

Many works in this area focus on social tags: Heymann (2008) proposed market-basket data mining to retrieve relevant tags. Agrawal (1993) used association rules that observe the relationships between tags from the co-occurrence relations of tags. Song (2011) proposed a general model of the description of tag expansion in a bigraph. Rae (2010) mentioned a computing mode for predicting another tag t based on some known tags. Its basic idea is that the probability of the known tags generating

---

[2] Our data are shared on http://xunren.thuir.org/share_EPSN/

tag *t* is the joint probability of each known tag producing tag *t*. This approach not only takes into account the conditional probability but also observes the prior probability *P(t)* of the predictive tags, which is in favor of recommending tags with high frequency and therefore helps to solve Inter-User disagreement. Schenkel (2008) presented a computing mode for expanding a tag *t* based on a known document. The probability of a tag being expanded is computed by the maximum probability of a certain keyword in the document. Liu (2009) put forward a random walk model over a tag graph to improve the ranking of tags. The idea is based on the probability that a tag of the object is related to all of the keywords in the graph. Li (2009) proposed a neighbor voting algorithm that accurately and efficiently learns tag relevance by accumulating votes from visual neighbors. They used 3.5 million tagged Flickr images and concluded that the voting method is very efficient and effective. Szomszor (2008) presented a method for automatic consolidation of users who are active in two social networks to have more tags to model user interests, which is also an important approach to tag expansion.

Importing friendship information to find users' private attributes has also been explored by many research works (Linda mood, 2009; Zheleva, 2009; Mislove, 2010). Certain experimental results show that friendship information can leak private information to some extent (Zheleva, 2009), while other results show that certain user attributes can be inferred with high accuracy when given information on as little as 20% of the users (Mislove, 2010). However, these works focus more on privacy protection and on general profiles such as location, grade in school, etc.; thus, models of these works are more related to community detection, analysis of networks, and privacy-related topics.

To summarize, many works inspired us to solve the problem of expanding self-tags, especially the works on social tagging. However, we believe that our paper is the first research effort to focus on expanding tags for linked users by using their self-tags and their friendship information and is also the first one to measure the performance of the algorithms on common users and celebrities separately and on different types of social relationships.

3   Problem Definition

We define a social network as a directed graph G (V, E, T), where V is a set of u nodes representing users in the social network, E is a set of following relations (the directed friendship links), T is a tag set of all users, and t (u) indicates self-tags that are viewed as a list of keywords u chosen from T. Finally, $G(u)$ is a subgraph of $G(V,E,T)$, where $V=\{u\}$, $E=\{$links related to $u\}$ and $T = t(u)$.

The problem of tag expansion of users in social networks can be generalized as solving the conditional probability of expanding tag e given the social networks G(u) and k self-tags$(t(u)_l,...t(u)_k)$.

$$P(e|\{t(u)_1, ... t(u)_k\}, G(u)), e \neq t(u)_1, ... t(u)_k$$

4 Our Work

4.1 Survey on the KL Divergences for Different Friendships

Traditional approaches such as joint probability and association rules mining only consider relations between tags or relations between users who tag the same content, while self-tags in social networks are given by users themselves instead of others, and as a result, there are no relations between users who tag the same users. Zheleva (2009) first proposed that user profiles can be inferred from their friendship and group information; therefore, we believe that tag expansion can also be inferred from users' relationship information, and the core of our survey methodology is to explore tags whose sources are the most similar to users' self-tags, their bidirectional following friends, followers, following sets or their own microblogs.

First, we define a basic function called *follow (u, x)*, which indicates the following relationship between *u* and *x*. If *u* follows *x*, then *follow(u,x)* = *true*; if not, then *follow(u,x)* = *false*. The social relationships of an objective user form a type of user set, and each user in this set has some relationship with the objective user. Based on the function of *follow*, we define 4 types of social relationships.

The Only-Following User Set of a user *u* (*OFS(u)*) contains the users followed by *u* instead of the users following *u*.

$$\text{OFS}(u) = \{x | follow(x, u) == false \land follow(u, x) = true\} \tag{1}$$

The Follower User Set of a user *u* (*FS(u)*) contains the followers of *u* instead of the users that *u* follows.

$$\text{FS}(u) = \{x | follow(x, u) == true \land follow(u, x) = false\} \tag{2}$$

The Bidirectional Following User Set of a user *u* (*BFS (u)*) contains the users following *u* and the users followed by *u*.

$$\text{BFS}(u) = \{x | follow(x, u) == true \land follow(u, x) = true\} \tag{3}$$

The ALL User Set of a user *u* (*ALL(u)*) contains the users who have at least a following link with *u*.

$$\text{ALL}(u) = \{x | follow(x, u) == true \lor follow(u, x) = true\} \tag{4}$$

We randomly select 0.32 million users who are divided into 5 ranges based on the number of their followers. Then, we observe the KL divergences between users' self-tags and tags from *OFS*, *FS*, and *BFS*. In *BFS*, for example, we extract all of the tags of users in BFS(*u*) and draw a probability table of the occurrence of all of the tags that can compute the probability *P(t)* of any of the tags in the table. The KL divergence formula is shown below:

$$\text{KL}(t(u), tag(BFS(u))) = -\sum_{i=1}^{k} \frac{1}{|t(u)|} \log\big(P(t_i)\big) \quad t_i \in t(u) \tag{5}$$

The experimental results of Table 1 show that KL divergences between users' self-tags and tags of *BFS* are the shortest, with an average KL divergence of 5.84.

The divergence between users' self-tags and tags of *FS* reaches 6.74, which is the largest. This result indicates that users share more tags with their bidirectional following friends than with their followers. On the other hand, common users who have fewer than 1000 followers are the mainstream crowd (78% are common users in the random sampling), and only 1% of the users are celebrities who have more than 1 million followers. The KL divergences are small for the crowd and large for celebrities, showing that the friendship of the crowd is simpler than that of celebrities who have diversified circles.

Table 1. KL divergences between user's self-tags and different types of social relationships

Followers-range	# users in the range	BFS	OFS	FS
[1M,∞]	146	5.8021	5.7265	7.6361
[100k ,1M]	3058	5.9733	6.2122	7.2468
[10k ,100k]	15636	5.9910	6.2834	6.8481
[1k,10k]	55016	6.0472	6.2288	6.4336
[0,1k]	266093	5.3895	6.0332	5.5275
Average		**5.8406**	**6.0968**	**6.7384**

In addition, to understand the relationship between tags and users' microblogs, we select 1000 microblogs of each user and divide these microblogs into words (removing stop words and other meaningless words) that act as tags. The results show that the KL divergences between the tags and the microblog contents of users are huge, with the smallest KL divergence of 6.44 in Table 2. The reason is that most users usually do not talk about content related to their tags, such as nationality, sex, educational background, profession, etc., in their microblogs but tend to discuss other non-privacy-related topics such as news and constellations. We plot all of the KL divergences in Figure 1.

Table 2. KL divergences between tags and microblogs of users

Followers-range	# users in the range	Blog
[1M,∞]	146	7.7720
[100k ,1M]	3058	7.2440
[10k ,100k]	15636	6.9016
[1k,10k]	55016	6.7707
[0,1k]	266093	6.4472

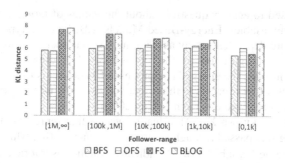

Fig. 1. KL divergences between tags of users and user sets of different types of social relationships

According to the survey in this section, we discover 4 basic facts:

- Tags of users and tags of their BFS have the maximum similarity.
- KL divergences between users' tags and the microblogs that they post are huge, especially for common users, who have no more than 1,000 followers.
- KL divergence greatly differs for different ranges of followers and is usually small for common users and large for celebrities.
- Common users are the mainstream users, and tag expansion should focus on common users who have fewer followers.

4.2   Our Baselines

As part of our research, we employ the approach of association rules mentioned by Heymann (2008) as our first baseline and consider the approach mentioned by Agrawal (1993) as our second baseline. Its formula is listed below:

$$P(e|t(u)) = P(e) * \prod_{t_i \in t(u)} \begin{cases} P(t_i|e), & if\ p(t_i|e) > 0 \\ \epsilon, & otherwise \end{cases} \qquad (6)$$

We compute the association probability of each tag e on users' self-tag set $t(u)$. The higher the probability, the greater relevance e has. We adopt the approach mentioned by Liu (2009) to give a good rank as our third baseline. First, obtain the *BFS* of a user; then, use the probability of tags of *BFS* as the prior (v_j) and the joint probability of tag i and tag j as p_{ij}; and finally, use the formula below:

$$r_k(j) = \alpha \sum_i r_{k-1}(i)p_{ij} + (1 - \alpha)v_j \qquad (7)$$

4.3   Our Model

Machine Learning usually can be viewed as a method to create a connection between X (known variables) and Y (target variables). By capturing such dependencies, a model can be used to answer questions about the values of target variables given the values of known variables. Energy-Based Models (EBMs) are a type of popular model that can capture dependencies by associating a scalar energy (a measure of compatibility) to each configuration of the variables and finding values of the target variables that minimize the energy. The process of Learning is to find an energy function that associates low energies with correct values of the target variables and higher energies with incorrect values (Yann LeCun 2006)

In this paper, the known X (u,t), a feature vector that represents the configuration of user u and tag t, is passed through a parametric function G_W, which produces a scalar output. The target variable $\hat{Y}(u, t)$ indicates whether user u regards tag t as a self-tag. The Energy function is the quadratic value of the difference between $G_W(X(u, t))$ and $\hat{Y}(u, t)$.

$$E\big(W, \hat{Y}(u, t), X(u, t)\big) = (G_W\big(X(u, t)\big) - \hat{Y}(u, t))^2 \qquad (8)$$

where $\hat{Y}(u, t) = \begin{cases} 1 & t \in tagset(u) \\ 0 & other\ wise \end{cases}$, tagset(u) is the set of self-tags of user u. The method of choosing the loss function is not the focus of our paper, so we just use negative log-likelihood loss, which works well in many architectures, as our loss function, and we omit u and t for conciseness

$$L(W, Y, X) = E(W, Y, X) + \frac{1}{\beta}\log\left(\int_{y \in Y} e^{-\beta E(W, y, X)}\right) \qquad (9)$$

It is natural to compute the gradient for each record $<X^i, Y^i>$ in the corpus and generate update rules as follows:

$$\frac{\partial L(W, Y^i, X^i)}{\partial W_x} = \frac{\partial E(W, Y^i, X^i)}{\partial W_x} - \cdot \sum_{y \in Y} \frac{\partial E(W, y, X^i)}{\partial W_x} * P(y | X^i, W) \qquad (10)$$

$$W_x \leftarrow W_x - \eta \frac{\partial L(W, Y^i, X^i)}{\partial W_x} \qquad (11)$$

Finally, we introduce the effective features of X(u,t) that we adopt in practice

- Prior probability of expanded tag t: $P(t)$
- Probability of expanded tag t generated by users' social graph $G(u)$: $P(t | G(u))$
- Probability of self-tag st_i, given expanded tag t: $P(st_i(u)|t)$

However, two obvious problems emerge:

- How do we construct the learning corpus?

There is a trick to constructing the learning corpus. First, we list each $P(t_i(u)|e)$ in the order of descending probability, say, for example, a user tags himself with A, B and C. Then, we construct a learning record of the expanded tag t. We just suppose that $P(B|e) > P(A|e) > p(C|e)$ and $t \in t(u)$, so the pair of learning records is then X(u,t) = { P(B|e),P(A|e),p(C|e),P(e|G(u)),P(e)} and $\hat{Y}$ (u,t) = 1.

In summary, we define the following features in Table 3 and sort the conditional probability of tag t given self-tags of user u in descending order:

Table 3. Features and their definition

Features	Definitions
1st-related	The largest conditional probability of self-tag given expanded tag t.
...	
nth-related	The smallest conditional probability of self-tag given expanded tag t.
G-power	Probability of social relationship G producing tag e
priori	Prior probability of tag e

- How do we choose G(u) and calculate P(t|G(u))?

We just simply calculate $P(t|G(u))$ as the frequency of the expanded tag t in all tags of users in $G(u)$; the reason for tf-idf-type approaches not being used is that the prior probability of tags has already been added to the regression calculations as a feature. Here, $G(u)$ can be replaced by different types of social relationships, such as *BFS, OFS,* and *FS,* as defined in Section 4.

5   Experiments

5.1   Datasets and Tools

We have launched and led a crowdsourcing organization and crawled 0.25 billion users' profiles, including name, sex, tags, introduction, verification, mutual following relationships, as well as their microblog contents of over 15 billion since 2010. As a result, we can obtain the data of the users' following and followers, together with the contents of their microblogs posted in Sina Weibo. In our experiments, we choose 10,000 users and their friend links as a training set, and 320,000 users and their 65 million following links and 4 billion followers (including duplicate followers of different users) as our test set.

We adopt *THUIRDB(Liang 2013)*, which has a good performance of completing millions of queries per second, as our database, which can effectively help our computing by indexing the following user sets and the followers' user sets.

5.2   Research Questions

We would first like to propose two main research questions in this paper and then carry out our experiments and analysis with these questions in mind.

1) Will the performance of tag expansion improve after importing friend information, and what are the differences between the performances of tag expansion based on different types of friend information?

2) Is the performance improvement effective on both common users and celebrities?

5.3   Training

We randomly choose 10,000 users as our training user set, from which we generate 20-40M learning materials as our training set. For our model, all of the features are normalized; therefore, the weight associated with each feature can reflect the importance of the feature to some degree. After training with the *SGD* algorithm, we give the weight of each feature in Table 4.

Table 4. Weight of features

Friendship type	Intercept	1st -R	2nd -R	...	G	priori
BFS	-6.24	4.21	1.93		19.51	0.41
OFS	-6.04	3.67	2.18		13.24	11.03
FS	-5.88	4.13	2.07		13.77	5.09
ALL	-6.55	3.80	2.24		13.08	8.42

Table 4 shows that friend information exerts great influence on the end results because the weight of the feature is the largest among all of the features. Moreover, some relevant tags have certain weights, reflecting the effect of the conditional probability of these users' self-tags. Feature *1st-R* being greatly larger than feature *2nd-R* in most cases shows that the most relevant user self-tags exert more influence than the other tags. Case studies show that users' self-tags are usually diversified, while expanded tags are normally only relevant to 1 to 2 users' self-tags; therefore, this result is also in line with our case studies.

5.4   Evaluation and Analysis

We randomly choose 0.32 million users with 10 tags of themselves among 0.25 billion users as our Test Set. Then, we randomly hide 5 self-tags of each user and expand tags based on the rest of the tags and their friendship information. Then, we compare these expanded tags with hidden tags to observe the *Precision* and *Recall*. For a clear description, we list the algorithms in Table 5. To make the description convenient, we will abbreviate *RW+BFS* algorithms to *RW* as one of our baselines, *BT+BFS* into *BFS*, etc., in the rest of the paper.

First, we would like to answer the first research question: will the performances of tag expansion improve after importing friend information, and what are the differences between the performances of tag expansion based on different types of friend information?

Figure 2 plots the results of this experiment for all evaluated users. Bars are plotted for each algorithm, and height is with respect to the value of *precision* or *recall*. Three important results can be observed in this graph. First, we note that the *BFS* algorithm outperforms the best baseline algorithm by over 14.0% on *P@1*, 14.4% on *P@5* and 11.4% on *R@10*. In fact, considering our large test set (0.32 million users and 1.6 million tag comparisons), the significance of our result is reasonable. Second, it is reasonable that the winner is *BFS*; as we observed in Section 4, the quality of *BFS* is the best source, i.e., tags from *BFS* have the shortest *KL* divergence to users' self-tags.

When we import all of the friend information, the performance of *ALL* becomes much worse than that of *BFS*. Overall, this experiment shows that users' self-tags can be effectively inferred by friendship links, especially bidirectional following friendship links and their self-tags.

Table 5. Algorithms used in evaluation

Algorithm	Description
AR	Association Rules (Hemann,2008)
JP	Joint Probability (Rae,2010)
RW+BFS(RW)	Random Walk by using *BFS* (Liu,2009)
BT+BFS(BFS)	our algorithm by using *BFS*
BT +OFS(OFS)	our algorithm by using *OFS*
BT+FS(FS)	our algorithm by using *FS*
BT+ALL(ALL)	our algorithm by using all social relationships

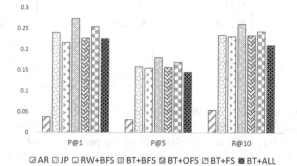

Fig. 2. Evaluations on all users

Additionally, Figure 3 plots the results of this experiment for the *Precision and Recall Curve* with each point (x=$P@k$, y=$R@k$ | k=1, 2, 3, 4, 5), and then sets line properties that make the baselines look like dashed lines and our algorithms look like solid lines. This result is convincing because we usually limit the windows of expanded tags in practice; therefore, if k output windows are available, the performances of $P@k$ and $R@k$ are of great importance. We observe a more significant tendency in Figure 3, specifically that our algorithms are capable of expanding tags for various windows. From the perspectives of *Average Recall*, *Average Precision* and *F-Score* shown in Table 6, we can tell that *BFS* is also the best one.

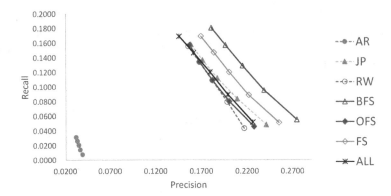

Fig. 3. Precision and Recall Curve

Table 6. Average Recall, Precision and F-Score

Algorithms	Average Recall	Average Precision	F-Score
AR	0.034	0.020	0.025
JP	0.192	0.107	0.138
RW	0.183	0.104	0.132
BFS	**0.220**	**0.123**	**0.158**
OFS	0.186	0.105	0.134
FS	0.205	0.115	0.147
ALL	0.206	0.115	0.148

To summarize our results for the first research question, performances of tag expansion have really been improved by importing friend information, whether viewed from the indexes *P@1, P@5* and *R@10* or from the perspective of *Precision and Recall Curve*, and the performance improves greatly, especially for *BFS* and *FS*. Other types of friend information are not as good as we imagined (they usually have a lot of noise), which is also consistent with our previous survey.

Next, let us come back to our second research question: is the improvement of performances effective for both common users and celebrities? We deliberately explore these results from the perspectives of P@1, P@5, and R@10.

P@1 is quite an important measurement because the performance of the best expanded tags usually represents the effectiveness of the algorithms on tag expansion. Figure 4 shows that the performance of the JP algorithm is the best baseline and that the BFS algorithm performs better than the best baseline algorithms in most cases. Case studies show that the tags of users with a large number of followers, e.g., invest and stock, tend to be subject-matter experts and hence lend themselves easily to association mining, while the tags of common users, e.g., Music, Runner, and Basketball, are usually high-frequency diversified words that are unlikely to have high-quality association rules; therefore, the performance of association mining will not be good enough. As for BFS, we discover that most common users with less than 1,000 followers are of high quality because they come from either the same school or the same company and share many similar tags. The performance of tag expansion (BFS) of celebrities with more than 1,000,000 followers is also great, due to their bidirectional following friends having a close background and identification, which is often reflected in the friends' tags. For users with a number of followers in the middle, i.e., between that of common users and celebrities, the performance is relatively poor due to the diversity of social relationships. Case studies show that these users are usually journalists, politicians and social activists.

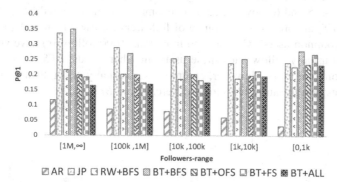

Fig. 4. *P@1* for different algorithms within different followers-ranges

Because we hide 5 tags, theoretically, if the expanded 5 tags are completely identical with the 5 hidden tags, then *P@5* may reach 100%. The experimental results in Figure 5 show that the *BFS* algorithm outperforms the baseline algorithms when the number of followers is smaller than 100,000. The result also shows that friendship information of common users is more effective, which is consistent with our survey in Section 4. For the *FS* algorithm shown by the green bar, due to the great differences in identification and background information between celebrities and their large number of followers, the performance of tag expansion is poor. However, identification and background information between common users are close to each other; hence, they share many similar tags, and the performance of the *FS* algorithm (green bar) becomes better and better with the reduction of the number of followers in Figure 5, which is in agreement with the findings of our survey in Table 1 in Section 4.

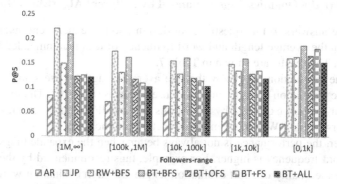

Fig. 5. P@5 for different algorithms within different followers-ranges

R@10 is also an important measurement to observe how much the rate of tags we hide can be recalled by each algorithm. The results in Figure 5 show that *BFS* can significantly outperform the baseline algorithms for common users. However, for celebrities, the performances of *JP* and *BFS* are similar. Because our experiment hides 5 of 10 tags, the *R@10* can be 50% at maximum; in fact, *R@10* of *BFS* and *JP* is more than 30% for celebrities, indicating that these algorithms can recall more than 3 tags that we have hidden before by generating 10 expanded tags. It is worth noting that Figures 4, 5 and 6 show the same phenomenon in that the red bar (*BFS*) and the green bar (*FS*) are similar in the range of followers of fewer than 1000. Case studies show that common users (who have no more than 1000 followers) have very few followers who usually follow back these users; in other words, *BFS* equals *FS* in most cases for common users. However, common users usually also follow a large amount of celebrities, which causes *OFS* to be different from *BFS* and *FS*.

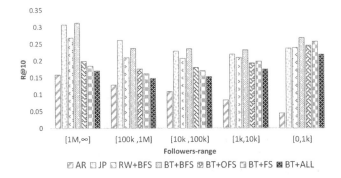

Fig. 6. R@10 for different algorithms within different followers-ranges

6   Discussion

After seeing the experimental results, 3 questions appear, which we discuss in depth in this section.

- **What Are the Qualities Tags Expanded by Different Algorithms Have?**

To find the answers to this question, we design another experiment and analyze the statistics on the average length and word frequency of tags recommended by different algorithms. The results are shown in Table 7.

After the calculation, we know that the average length of users' tags is 2.918758 and that the evaluation word frequency of users' tags is 0.036797. The results of Table 8 show that the tags expanded by algorithms AR and JP are most identical to the real tags of users, while RW and our methods prefer more hot tags.

However, the performance is not always better when the expanded tags are shorter and the word frequency is higher. For example, tags recommended by the ALL algorithm have the quality of short evaluation length and high evaluation word frequency but with poor performance. After taking following and followers into consideration, we find that popular tags are more easily outstanding, thus diluting the proportion of tags more related to users. This indicates that being popular is not necessarily good, but it is important to have high relevance.

Table 7. Word frequency and length of tags

Algorithms	Average length	Average word frequency
AR	3.04328	0.00011
JP	2.62578	0.02907
RW+BFS	2.15382	0.11213
BT+BFS	2.11390	0.12047
BT+OFS	2.00250	0.12824
BT+FS	2.02436	0.12437
BT+All	2.00150	0.13100

- **What will Happen If We Only Adopt the Feature of Social Networks?**

In our algorithms, we take into account the prior probability of the expanded tags and the relevance between expanded tags and users' real tags. What will happen if we only consider the tag weight of social networks, i.e., expanding tags on the tags appearing most frequently in users' social relationships, instead of considering the two factors mentioned above? To find the answer to this question, we design a comparison experiment using the data of all users without dividing the range of followers. The result is shown in the following table:

Table 8. Performance of algorithms only considering the feature of social networks

Algorithms	P@1	P@5	R@10
AR	0.038	0.031	0.054
JP	0.240	0.158	0.234
BT+BFS	0.273	0.180	0.260
BT+BFS_ONLYG	0.211	0.070	0.111
BT+OFS	0.226	0.156	0.233
BT+OFS_ONLYG	0.167	0.059	0.099
BT+FS	0.254	0.169	0.243
BT+FS_ONLYG	0.200	0.067	0.107
BT+ALL	0.225	0.145	0.210
BT+ALL_ONLYG	0.193	0.065	0.104

Algorithms with the suffix ONLYG represent algorithms only considering the feature of social networks instead of other features. We discover that under this circumstance, the four main indexes decrease in an obvious fashion. For example, in BFS, P@1 decreases by 22%, P@5 decreases by 61%, and R@10 decreases by 57%.

This indicates that without considering the relevance between expanded tags and us-ers' real tags, the performance of words with high frequency on social networks will decrease greatly.

- **Does Our Algorithms Perform Better Only If the Number of Followers Is Small?**

We suppose that this phenomenon may be related to the fact that the subjects of train-ing samples in the process of training feature weights are users with a small number of followers. Therefore, we train users with over 10k followers to learn another set of parameters and observe the change in the performance. We only use the BFS algo-rithm to observe the trend.

Table 9. Weights of features for training of different users

	All users	Users with >10K followers
Intercept	-6.2415	-6.1342
1st –R	4.2188	4.4144
2nd -R	1.9302	2.3467
3rd -R	1.2862	0.5856
4th -R	0.2528	1.8256
5th -R	2.6454	2.0146
6th -R	0.9987	0.6409
7th -R	1.3642	1.0403
8th -R	3.2611	10.484
9th -R	-7.3255	-11.045
10th -R	-62.705	-117.535
G-power	19.5	27.7
priori	0.4121	-7.5465

From Table 9, we discover that the importance of the prior of the expanded tags is weakened and that the effect of the social networks is stronger when carrying out pa-rameter training with users who have over 10k followers.

Table 10. The improvement by using training set of users with over 10k followers

Followers-range	BT+BFS			BT+BFS-10K		
	P@1	**P@5**	**R@10**	**P@1**	**P@5**	**R@10**
[1M,∞]	0.347	0.207	0.312	0.372	0.213	0.318
[100k ,1M]	0.270	0.161	0.237	0.283	0.173	0.255
[10k ,100k]	0.256	0.156	0.229	0.276	0.172	0.252
[1k,10k]	0.247	0.153	0.224	0.262	0.166	0.244
[0,1k]	0.265	0.175	0.252	0.273	0.180	0.262

We design the same experiment on this set of parameters and name it BFS-10K. The results in Table 10 show that nearly every measurement improves, especially when the number of followers is over 10k, indicating a great improvement in the performance after changing the training corpus. This fully indicates that the performance of expanded tags still has room for improvement and is closer to the real data when we use users with different ranges of followers for learning.

Finally, after some deep discussions, we draw several important conclusions:

- Tags expanded by our algorithms have relatively high word frequency and short length.
- If we only take into account the weight of social networks without considering the relevancy of the tags that users already have, the performance will decrease greatly.
- After using a new training set with users having over 10k followers on our algorithms, we discover that the performance of our algorithms increases greatly.

7   Conclusion

This paper puts forward and defines the problem of expansion of self-tags of users in social networks. Under our definitions of four types of social relationships (BFS, OFS, FS and ALL), we discover that users' tags are more similar to the tags of their bidirectional following friends (BFS). We only choose as features the tag frequencies of users' social relationships, the conditional probabilities of friends' tags given users' self-tags and the prior probabilities of tags and adopt energy-based learning to build a model and use a negative log-likelihood loss as the loss function. The experimental results indicate that our algorithm outperforms the best baseline algorithm by over 14.0% on P@1, 14.4% on P@5 and 11.4% on R@10. Moreover, experiments also show that BFS is better for celebrities and that BFS as well as FS is better for common users.

Compared with traditional methods, our method inherits the previous work and imports friendship information into the modeling to gain an obvious improvement, which encourages us to go further in this direction. Future work will include the following: 1) exploration of more complex algorithms; 2) prediction of users' expanded windows; 3) cross-social network tag expansion; and 4) expansion of tags from the contents of microblogs of users' friends.

Acknowledgments. This work was supported by Natural Science Foundation (60903107, 61073071), National High Technology Research and Development (863) Program (2011AA01A205) and Research Fund for the Doctoral Program of Higher Education of China (20090002120005).

References

1. Agrawal, R., Imieliński, T., Swami, A.: Mining association rules between sets of items in large databases. ACM SIGMOD Record 22(2), 207–216 (1993)
2. Bernstein, M., Tan, D., Smith, G., et al.: Collabio: A game for annotating people within social networks. In: Proceedings of the 22nd Annual ACM Symposium on User Interface Software and Technology, pp. 97–100. ACM (2009)
3. Bowman, S., Willis, C.: We media: How audiences are shaping the future of news and information (2003)
4. Eck, D., Lamere, P., Bertin-Mahieux, T., Green, S.: Automatic generation of social tags for music recommendation. In: Advances in Neural Information Processing Systems, pp. 385–392 (2007)
5. Farrell, S., Lau, T., Nusser, S., et al.: Socially augmenting employee profiles with people-tagging. In: Proceedings of the 20th Annual ACM Symposium on User Interface Software and Technology, pp. 91–100. ACM (2007)
6. Goel, A., Gupta, P., Lin, J., et al.: Wtf: The who to follow service at twitter. In: Proceedings of the 22nd International conference on World Wide Web, pp. 505–514 (2013)
7. Garg, N., Weber, I.: Personalized tag suggestion for flickr. In: Proceedings of the 17th International Conference on World Wide Web, pp. 1063–1064. ACM (2008)
8. Heymann, P., Ramage, D., Garcia-Molina, H.: Social tag prediction. In: Proceedings of the 31st Annual International ACM SIGIR Conference on Research and Development in Information Retrieval, pp. 531–538. ACM (2008)
9. Jäschke, R., Marinho, L., Hotho, A., Schmidt-Thieme, L., Stumme, G.: Tag recommendations in social bookmarking systems. AI Communications 21(4), 231–247 (2008)
10. Kucuktunc, O., Sevil, S.G., Tosun, A.B., Zitouni, H., Duygulu, P., Can, F.: Tag suggestr: Automatic photo tag expansion using visual information for photo sharing websites. In: Duke, D., Hardman, L., Hauptmann, A., Paulus, D., Staab, S. (eds.) SAMT 2008. LNCS, vol. 5392, pp. 61–73. Springer, Heidelberg (2008)
11. Kullback, S., Leibler, R.A.: On information and sufficiency. The Annals of Mathematical Statistics 22(1), 79–86 (1951)
12. LeCun, Y., Chopra, S., Hadsell, R., et al.: A tutorial on energy-based learning. Predicting Structured Data (2006)
13. Liang, B., Liu, Y., Zhang, M., Ma, S., Zhang, K.: Predicting Tags for None-tagged Person on SNS. Journal of Computational Information Systems 10(8), 3123–3132
14. Liang, B., Liu, Y., Zhang, M., Ma, S.: THUIRDB: A large-scale, highly-efficient index, fast-access key-value store. Journal of Computational Information Systems 9(6), 2347–2355 (2013)
15. Li, X., Snoek, C.G., Worring, M.: Learning social tag relevance by neighbor voting. IEEE Transactions on Multimedia 11(7), 1310–1322 (2009)
16. Lindamood, J., Heatherly, R., Kantarcioglu, M., et al.: Inferring private information using social network data. In: Proceedings of the 18th International Conference on World Wide Web, pp. 1145–1146. ACM (2009)
17. Liu, D., Hua, X.S., Yang, L., Wang, M., Zhang, H.J.: Tag ranking. In: Proceedings of the 18th International Conference on World Wide Web, pp. 351–360. ACM (2009)
18. Mislove, A., Viswanath, B., Gummadi, K.P., et al.: You are who you know: Inferring user profiles in online social networks. In: Proceedings of the Third ACM International Conference on Web Search and Data Mining, pp. 251–260. ACM (2010)
19. McPherson, M., Smith-Lovin, L., Cook, J.M.: Birds of a feather: Homophily in social networks. Annual Review of Sociology 27(1), 415–444 (2001)

20. Muller, M.J., Ehrlich, K., Farrell, S.: Social tagging and self-tagging for impression management. Submitted as an Interactive Poster to CSCW (2006)
21. Chodorow, K.: MongoDB: The definitive guide. O'Reilly Media, Inc. (2013)
22. Rae, A., Sigurbjörnsson, B., van Zwol, R.: Improving tag recommendation using social networks. In: Adaptivity, Personalization and Fusion of Heterogeneous Information, pp. 92–99. Le Centre de Hautes Etudes Internationales D'Informatique Documentaire (2010)
23. Schenkel, R., Crecelius, T., Kacimi, M., Michel, S., Neumann, T., Parreira, J.X., Weikum, G.: Efficient top-k querying over social-tagging networks. In: Proceedings of the 31st Annual International ACM SIGIR Conference on Research and Development in Information Retrieval, pp. 523–530. ACM (2008)
24. Song, Y., Zhang, L., Giles, C.L.: Automatic tag recommendation algorithms for social recommender systems. ACM Transactions on the Web (TWEB) 5(1), 4 (2011)
25. Sood, S., Owsley, S., Hammond, K.J., et al.: TagAssist: Automatic Tag Suggestion for Blog Posts. In: ICWSM 2007 (2007)
26. Szomszor, M., Alani, H., Cantador, I., O'Hara, K., Shadbolt, N.: Semantic modelling of user interests based on cross-folksonomy analysis. In: Sheth, A.P., Staab, S., Dean, M., Paolucci, M., Maynard, D., Finin, T., Thirunarayan, K. (eds.) ISWC 2008. LNCS, vol. 5318, pp. 632–648. Springer, Heidelberg (2008)
27. Wal, T.V.: Folksonomy coinage and definition (2007), http://www.vanderwal.net/folksonnomy.html
28. Wal, T.V.: Explaining and showing broad and narrow folksonomies (2005), http://wwww.vanderwal.net
29. Weng, J., et al.: Twitterrank: finding topic-sensitive influential twitterers. In: Proceedings of the Third ACM International Conference on Web Search and Data Mining. ACM (2010)
30. Zheleva, E., Getoor, L.: To join or not to join: The illusion of privacy in social networks with mixed public and private user profiles. In: Proceedings of the 18th International Conference on World Wide Web, pp. 531–540. ACM (2009)

Predicting the Popularity of Messages on Micro-blog Services

Yang Li, Yiheng Chen, Ting Liu, and Wenchao Deng

Research Center for Social Computing and Information Retrieval,
Harbin Institute of Technology, Harbin, China
{yli,yhchen,tliu,wcdeng}@ir.hit.edu.cn

Abstract. Micro-blogging is one of the most popular social media services on which users can publish new messages (usually called tweets), submit their comments and retweet their followees' messages. It is retweeting behavior that leading the information diffusion in a faster way. However, why some tweets are more popular than others? Whether a message will be popular in the future? These problems have attracted great attention. In this paper, we focus on predicting the popularity of a tweet on Weibo, a famous micro-blogging service in China. It is important for tremendous tasks such as breaking news detection, personalized message recommendation, advertisement placement, viral marketing etc. We propose a novel approach to predict the retweet count of a tweet by finding top-k similar tweets published by the same author. To find the *top-k* similar tweets we consider both content similarity and temporal similarity. Meanwhile, we also integrate our method into a classical classification method and prove our method can improve the results significantly.

Keywords: micro-blogs, popularity, retweet count, similarity.

1 Introduction

In recent years, Micro-blog Services are developing rapidly, such as Twitter[1] and Weibo[2]. Users can publish their information through SMS, instant message, email, web sites or third-party applications by inputting short text with a constraint of 140 characters, which are called tweets or statuses [1]. One important behavior on microblogs is retweeting. It allows users to share information with their followers and the information spreads more widely via the social network. Micro-blog services have become a new medium for information propagation not just a social network [2].

Study of information diffusion in microblogs has gained wide attention, among which one basic problem is to predict the future popularity of a tweet. As the retweet count of a tweet depends on the quality of its content. The popularity of a tweet is usually represented by the retweet count. Predicting future retweet counts for individual tweets provides one solution to the problem of information overload, since it can help recognize high-quality and potentially popular tweets in advance from massive

---

[1] http://twitter.com
[2] http://www.weibo.com

H.-Y. Huang et al. (Eds.): SMP 2014, CCIS 489, pp. 44–54, 2014.

collections of user-generated content. It is also very important to a number of tasks such as hot topics detection, viral marketing and public opinion monitoring, and so on.

Yang et al. used a regression model to predict the speed, scale and range of diffusion on some topics but get a poor result [3]. Hong et al. casted the problem into classification method to predict the volume of retweets a particular message will receive in the near future [4]. Bakshy et al. demonstrated that everyone is an influencer and has an general influence over retweet count [5]. They proposed to use average retweet count of one's past tweets to predict for the current tweet published by the same author. However, they ignore the fact that for the same author, different content and different publishing time usually lead to different retweets.

Inspired by the earlier work, in this paper we propose a novel unsupervised approach to predict the popularity of a tweet based on finding top-k similar tweets that are published by the same author (called k-Sim_retweets approach). In our method, to predict the retweet count of a target tweet, we collect the tweets published by the same author in the past firstly. We call these tweets "*author-identical tweet set*". As in this set the main differences between tweets are the content and the publishing time. We compute the similarity between tweets in Euclidean distance by incorporating weights of content features and temporal features, and extract top-k similar tweets to the target one. At last, their weighted mean retweet count represents the estimated retweet count of the target tweet. We compared our method with many baseline methods on a ground truth dataset. Our approach achieves comparable results with supervised learning method and also can be integrated to improve the prediction performance.

The contribution of the paper can be summarized as follows:

- We propose a novel unsupervised prediction approach based on the idea of extracting top-k similar author-identical tweets for target tweet.
- To extract the top-k similar author-identical tweets, we investigate similarity computation on both content and temporal features.
- Our method is independent of other context features and can be integrated to classification solutions to improve the performance significantly.

2   Related Work

Predicting popularity of information is one of the major challenges in research of social networks. In order to know what kind of information people like to propagate on Twitter, several studies have been conducted by studying the retweet behavior of the users. Boyd et al. analyzed retweets from a conversational view and find how people retweet, why they retweet and what they retweet[6]. Suh et al. utilized PCA to study the basic factors of retweeting[7]. Recently, related to our work many researchers focused on examining the possibility of a tweet being forwarded., some treats the problem as a classification problem that is predicting whether a tweet can be retweeted in the future[1]. Others treat the problem as a regression problem that is predicting the information popularity[3]. Kupavskii et al. predicted how many retweets a tweet will gain in a fix time period by using an important feature flow of information cascade[8]. Our work is to predict retweet count of a tweet without knowing any progation information about it.

3   Dataset

To begin with, we first randomly crawled 100 seed users and all their user profiles which contain names, genders, verification statuses, numbers of bi-friends, numbers of followers, numbers of followees, and number of tweets. Then we crawled their tweets from January 1, 2013 to April 1, 2014 and filtered out whose followers and followees are less than 100 or whose tweets are less than 100. At last, the dataset is consisted of 83 users and 60,362 tweets. We split the dataset from August 1, 2013 into two parts. The preceding part includes the tweets before August 1, 2013 as the history data. The posterior part includes the tweets after October 1, 2013 as the target tweets to be predicted with a total number of 24,626 tweets. Table1 shows the statistics of our datasets. Sysomos, a company of social media monitoring and analytics solutions, analyzed 1.2 billion tweets in the year 2010 and found 92% retweets of a tweet happened within the first hour after it is published. We make statistics for retweet counts of tweets at least one month after the time it is published to make sure the tweet will not gain any more retweets. These are the gold standard of our experiments.

Table 1. Statistics of datasets

DataSet	#users	#tweets
Total	100	64,253
History Dataset	83	39,736
Experiment Dataset	83	24,626

4   Popularity Prediction

4.1   Problem Definition

Weibo is one of the most popular Chinese microblog services. Users can write short text no more than 140 characters and upload pictures, videos or audios to share information with their friends. Then his friends can retweet the information so that makes the information presented in front of a new set of users. Therefore retweet count is a direct way to evaluate the popularity of tweets.

In this paper, we define an instance in our prediction model as a tuple $<a, t, p>$, where a denotes the author of the tweet, and t denotes the tweet itself. We focus on predicting the popularity p for t at the initial time that the tweet is published. It is difficult to predict the exact number of retweets for a tweet. Instead, we divide the messages into different retweet volume "classes" similarly to [9]: 0: Not retweeted, 1: Retweeted less than 10, 2: Retweeted more than 10 and less than 100, 3: Retweeted more than 100 and less than 1000, 4: Retweeted more than 1000 and less than 10000, 5: Retweeted more than 10000. Figure1 shows the distribution of retweet counts in this dataset which follows the power-law distribution. The number of instances in every class is very similar. The six classes stand for different levels of popularity trends. Thus our main task is to predict which class the tweet will belong to.

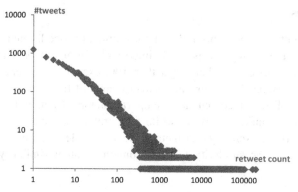

Fig. 1. The distribution of retweet counts in the dataset

4.2   Main Method

Every user has personal influence in information spreading. Bakshy et al.[5] demonstrated that everyone has an average influence over retweets count. They use average retweet count of a user as the prediction retweet count. But they ignored the factors of topics and time. For example, considering a user who is a famous expert on data mining, his tweet about Data Mining will attract many retweets relative to the tweets on Machine Vision. Considering on both content factor and temporal factor, we proposed an author-identical approach to predict the retweet number for a tweet by extracting the top-k similar tweets on a weighted feature model. The main method we proposed shows as Algorithm1.

Algorithm1

Input: t_{target} //Author a, Target tweet

$T<t_0, t_1, t_2, \ldots t_n>$ //Author-identical tweet set

K   // The number of selected tweets in the two tweet set

α,β //The weight of the features   and $\alpha+\beta=1$

Output: $RetweetCount\_t_{target}$ //The retweet number

Procedure:

1.   Calculate $T_{content\_sim}$: The content similarity between t and every tweet in T

2.   Calculate T_{time_sim} : The temporal similarity between t and every tweet in T

3.   Sort $T_{content\_sim}$ and T_{time_sim} by descending order.

4.   Calculate $RetweetCount\_t_{target}$ for t_{target} , where $t_i \in T_{content\_sim}$, $t'_i \in T_{time\_sim}$

$$RetweetCount\_t_{target} = \alpha \sum_{i=1}^{k} RetweetCount\_t_i + \beta \sum_{i=1}^{k} RetweetCount\_t'_i$$

4.3   Features

The retweet counts will probably fall in a conventional scope limited by the author's social influence. For example, it is nearly impossible for an author with less than 100 followers to get a retweet number larger than 100, except that a very famous person helps retweet the tweet. However, the retweet counts of the same author are usually different due to different content and temporal factors. Content feature represents for the internal semantic of a tweet while temporal feature carters for the external condition of the tweet. Actually, our method is an example based method to find most similar the tweets. The similarity features contains content similarity and temporal similarity.

Content Similarity
Noun phrases usually stand for the main semantic of a tweet in the research of text mining in microblogs [10]. Unlike other social network services, the text in a tweet is usually very short. Figure 2 show that the statistics of the length of tweets in our dataset. Although 140 words is the upper bound of the input text in our data set, but above 40% tweets contain less than 80 words. In the dataset, the average length of tweets is about 86 without urls, hashtags and symbols. Then for every tweet, the length of noun phrases dividing by the length of the whole sentence is adopted to get the proportion of noun phrases. In Figure 3 we found that the average proportion is larger than 30% which means noun phrases can almost represent the main idea of a tweet.

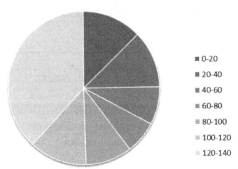

Fig. 2. The statistics of average length of tweets in the dataset

In order to calculate the content similarity, we first conduct word segmentation and part-of-speech tagging on the text of the target by using Language Technology Platform (LTP)3. Secondly, we use some rules based on part-of-speech to extract noun phrases in it. At last, the similarity between the tweets is computed. Our idea is to compare two texts by means of their noun phrases. Every tweet can be represented as an N-dimensional Vector. N is the total number of noun phrases. The value of each dimension is the amount of corresponding noun-phrase existed in the tweet. Then we calculate the cosine similarity between two tweets.

---

[3] http://www.ltp-cloud.com/

$$ContentSimilarity(\ x, y\ )= \frac{x_1 y_1 + x_2 y_2 + ... + x_n y_n}{\sqrt{x_1^2 + x_2^2 + ... + x_n^2}\sqrt{y_1^2 + y_2^2 + ... + y_n^2}} \tag{1}$$

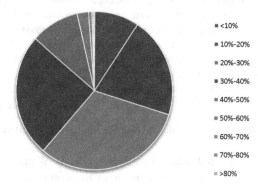

- <10%
- 10%-20%
- 20%-30%
- 30%-40%
- 40%-50%
- 50%-60%
- 60%-70%
- 70%-80%
- >80%

Fig. 3. The proportions of noun phrases for every tweet in the dataset

Temporal Similarity

Temporal factor is an important external factor of retweeting. In our dataset, a tweet published at mid-night will hardly get more retweets than in the morning due to the temporal difference. That means followers like to retweet more in the morning than at mid-night. In order to verify this hypothesis, we analyze the "follower active time" by the time of their publishing tweets. If they publish a tweet at a certain period of time, they are probably online. So for a target author, the more active his followers are the more retweets they will gain in a specific period of time. Figure 4 shows the tweets published by all the followers of the seed users on different time spans. We find the followers are more active at daytime which means they have greater probability of retweeting at daytime than at mid-night.

As stated above, temporal factor can affect retweeting behaviors. Tweets publishing on similar time period will probably get similar retweeting due to the similar activeness of the followers. However, for the same author, his influence is various by the time especially for the celebrities. For example, some negative news will lead to a decline in the number of followers.

So we consider the temporal difference in the dimensions of specific hour in a day and the specific date. The temporal similarity between two tweets is formulated as Formula 2, where y, m, w and h stands for different units of time(the year, month,

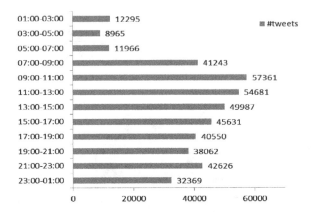

Fig. 4. The statistics of the tweets of the followers on different time spans

week and hours) in the published time of tweet, Y, M, W and H are the normalization factors. $\lambda 1$, $\lambda 2$, $\lambda 3$, $\lambda 4$ are the weights with the constraint $\lambda 1+\lambda 2+\lambda 3+\lambda 4=1$. In the following experiments, we set $\lambda 1=\lambda 2=\lambda 3=\lambda 4$.

$$TimeSimilarity(t_a,t_b)=\sqrt{\lambda_1\left(\frac{y_a-y_b}{Y}\right)^2+\lambda_2\left(\frac{m_a-m_b}{M}\right)^2+\lambda_3\left(\frac{w_a-w_b}{W}\right)^2+\lambda_4\left(\frac{\min(|h_a-h_b|,24-|h_a-h_b|)}{H}\right)^2} \quad (2)$$

5   Experiment

5.1   Parameter Settings

The idea of our method is to find *top-k* similar tweets on both content and temporal factors. We randomly sampled 10000 tweets in the experiment dataset to train the parameter k and the weights. Figure 5 shows the accuracy of the method on different k values and $\alpha=\beta=0.5$. That means content feature and temporal feature are equally important. When $k=4$ the method gets a best accuracy of 60.52%. The results are showed in Figure 4. When $k\geq4$, the accuracy begins to decline. Then we set the parameter k as 4 in the following experiments.

In order to see which factor is more important to retweet count, we use an EM liked algorithm to learn the weights α and β in the case of $k=4$. The method is to maximize the accuracy in the condition of $\alpha+\beta=1$ and step $=0.1$. Figure 6 shows the results of different settings of α and β. While $\alpha=0.7$ and $\beta=0.3$ we get a highest accuracy of 60.78%. When α equals 0 or β equals 0, it reaches a lowest point. That means both content and temporal factor are important to retweet count. The curve graph presents the accuracy of the method is changing smoothly with α and stays stable at around 60%. It makes no much differences when α is between 0.2 and 0.8. Therefore, in the following experiments, we set $k=4$, $\alpha=\beta=0.5$.

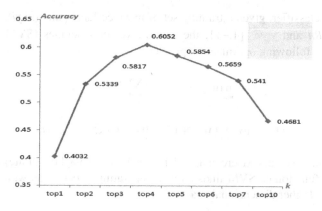

Fig. 5. Accuracy of the method with different k values and $\alpha=\beta=0.5$ on the sampled dataset

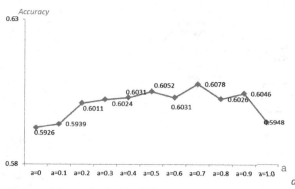

Fig. 6. Accuracy of the method with different α values on the sampled dataset

5.2   Classification Framework

The method we proposed is unsupervised and only uses content and publishing time of tweets. Then we integrate our method to a supervised approach similar to [4] to show the effectiveness of the method and its scalability. Figure 6 shows our feature based classification framework. For every training tweet, we first perform our method to get a preliminary result by computing the top-k similar tweets' weighted retweet count. Then use the result as an additional feature together with other classical features to train the prediction model. Actually we use almost all features in Hong' work [4]. The features are listed in Table 2. Finally we can use the model to predict the retweet count.

Specially, we use a support vector machines (SVM) model to learn a multiple classification framework by means of "One versus rest, OVR". Create k binary classifier (k is the number of classes). Among them, the i-th classifier seprates the instances belonging to i-th class from the rest. We treat all other instances as negative class and focus on the i-th class while we are training for the i-th classifier.

For every classifier, given a training set of instance-label pairs (xi, yi), $i = 1,...,l$, where $xi \in Rn$ and $y \in \{1, -1\}$, the support vector machines (SVM) require the solution of the following optimization problem:

$$\min_{w,b,\xi} \frac{1}{2} w^T w + C \sum_{i=1}^{l} \xi_i \quad (3)$$

$$\text{Subject to} \quad y_i(w^T \phi(x_i) + b) \geq 1 - \xi_i, \quad \xi_i \geq 0 \quad (4)$$

Here training vectors xi are mapped into a higher (maybe infinite) dimensional space by the function φ. SVM finds a linear separating hyper plane with the maximal margin in this higher dimensional space.

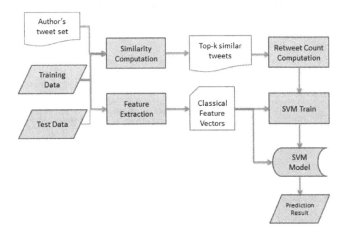

Fig. 7. Classification Approach for Retweet Count Prediction

Table 2. Features used in SVM-based classification framework

Author-based Features	*number of followers, number of followees, number of bi-friends, number of tweets, register time, is verified*
Tweet-based Features	*length, number of urls, number of hashtags, number of mentions, whether contains picture, similarity between tweets and user profiles*

5.3   Results

We perform several baseline methods to compare with our method. In this part we divide the rest of the experiment dataset into training and test data, and perform 5-fold cross validation. Table 3 shows the prediction performance evaluated on precision, recall and F-score. First, we directly use author's average retweet count to predict retweet count of a new tweet referred to [5]. The result is worst. Excitingly by using k-Sim_retweets only, we get a much better F-score of 57.41%. This is because average retweet count stands for the general influence of the author. They ignore the fact that different content and different time usually lead to different attention. Our method (k-Sim_retweets) takes both content and time into consideration on finding the top-k similar tweets.

Table 3. Comparision of Results

Method	Precision	Recall	F-score
avg_retweets	51.72%	54.39%	53.02%
k-Sim_retweets	62.58%	53.03%	57.41%
Hong 2011	61.09%	**59.29%**	60.18%
Hong 2011+ avg_retweets	59.94%	58.59%	59.26%
Hong 2011+ k-Sim_retweets	**64.93%**	58.69%	**61.65%**

Specially, k-Sim_retweets approach can get a comparable result with the supervised classification method (Hong2011) except for the recall. Although the classification method can get a better result than k-Sim_retweets method, it takes much more effort to extract various features and to train the model. Our method is easier to perform and can be integrated to the classification model as an additional feature. Finally we get an F-score of 61.65% and it outperforms the other two baseline methods (*Hong2011* and *Hong2011+ avg_retweets*).

6   Conclusion

In this paper, we propose a novel approach to predict the popularity of a tweet based on extracting the top-k similar tweets in a feature weighted model. The top-k similar candidates to the target one are selected from the *author-identical tweet set*, and their weighted mean of count of retweets is calculated as the estimated retweet count of the target tweet. Our method is much easier than other classification methods and achieves a good result. Above all, our method improves the performance when integrated to a state of art classification solution.

However, our work still suffers from some limitations. To find the top-k similar tweets, we compute similarities on temporal and content factors. First, when a tweet contains some new noun phrases, it is hard to find similar tweets based on content similarity. Then time will be the main factor of retweets which may lead the results incorrect. Second, the computation method we use is simple. We will analyze more factors and improve the method. Besides, we will try other parameter learning method such as genetic algorithm in our future work.

References

1. Petrovic, S., Osborne, M., Lavrenko, V.: RT to Win! Predicting Message Propagation in Twitter. In: ICWSM (2011)
2. Kwak, H., et al.: What is Twitter, a social network or a news media? In: Proceedings of the 19th International Conference on World Wide Web. ACM (2010)
3. Yang, J., Counts, S.: Predicting the Speed, Scale, and Range of Information Diffusion in Twitter. In: ICWSM (2010)
4. Hong, L., Dan, O., Davison, B.D.: Predicting popular messages in twitter. In: Proceedings of the 20th International Conference Companion on World Wide Web. ACM (2011)
5. Bakshy, E., et al.: Everyone's an influencer: Quantifying influence on twitter. In: Proceedings of the Fourth ACM International Conference on Web Search and Data Mining. ACM (2011)
6. Boyd, D., Golder, S., Lotan, G.: Tweet, tweet, retweet: Conversational aspects of retweeting on twitter. In: 2010 43rd Hawaii International Conference on System Sciences (HICSS). IEEE (2010)
7. Suh, B., et al.: Want to be retweeted? large scale analytics on factors impacting retweet in twitter network. In: 2010 IEEE Second International Conference on Social Computing (Socialcom). IEEE (2010)
8. Kupavskii, A., et al.: Prediction of retweet cascade size over time. In: Proceedings of the 21st ACM International Conference on Information and Knowledge Management. ACM (2012)
9. Khabiri, E., Hsu, C.-F., Caverlee, J.: Analyzing and Predicting Community Preference of Socially Generated Metadata: A Case Study on Comments in the Digg Community. In: ICWSM (2009)
10. Schumaker, R.P., Chen, H.: Textual analysis of stock market prediction using breaking financial news: The AZFin text system. ACM Transactions on Information Systems (TOIS) 27(2), 12 (2009)

Detecting Anomalies in Microblogging via Nonnegative Matrix Tri-Factorization

Guowei Shen, Wu Yang, Wei Wang, Miao Yu, and Guozhong Dong

Information Security Research Center, Harbin Engineering University, Harbin,
Heilongjiang Province, China 150001
{shenguowei,yangwu,w_wei,yumiao,dongguozhong}@hrbeu.edu.cn

Abstract. With the increasing of anomalous user's intelligent, it is difficult to detect the anomalous users and messages in microblogging. Most of the studies attempt to detect anomalous users or messages individually nowadays. In this paper, we propose a co-clustering algorithm based on nonnegative matrix tri-factorization to detect anomalous users and messages simultaneously. A bipartite graph between user and message is built to model the homogeneous and heterogeneous interactions, and homogeneous relations as constraints to improve the accuracy of heterogeneous co-clustering algorithm. The experimental results show that the proposed algorithm can detect anomalous users and messages with high accuracy on Sina Weibo dataset.

Keywords: Microblogging, Anomaly detection, Nonnegative matrix tri-factorization, User interaction behaviors.

1 Introduction

Due to the number of user growth and rapid information diffusion, microblogging has attracted significant attention recently. Many microblogging service platforms like Twitter and Sina Weibo[1] provide microblogging services to post messages, which called tweets. There are many researches on Twitter nowadays, while as the most popular microblogging platform in China, Sina Weibo has not been further researched yet. User behavior and content of message in Sina Weibo are very different from Twitter. Billions of messages are posted in Sina Weibo every day and propagated quickly through user's interaction behaviors. However, some users generated anomalous interaction behaviors or anomalous tweets for some special purposes. How to detect anomalous users and messages is essential for the social networking security and recommendation.

Traditional methods of anomalous detecting for users and messages are processed individually, but spammers become more and more intelligent. Some spammers usually post normal messages, only occasionally post some promotional activities, advertising, spam messages and so on. Only use the attribute of user as the judgment condition can lead to miss some anomalous users. Similarly, if only use the content of

---

[1] http://weibo.com/

H.-Y. Huang et al. (Eds.): SMP 2014, CCIS 489, pp. 55–66, 2014.

the message as the judgment condition, the accuracy may be very low. Therefore, user and message are mutually related, while user and message are considered simultaneously, the accuracy can be improved, and reducing processing time.

In this paper, we propose a co-clustering algorithm based on nonnegative matrix tri-factorization to detect anomalous users and messages simultaneously. Through the analysis of the relationship extracted from Sina Weibo, both user and message are important entities in microblogging, so a bipartite graph between user and message is constructed to model homogeneous and heterogeneous interactions.

This paper focuses on anomaly detecting based on the user interaction behavior in Sina Weibo, and makes the following major contributions:

- A bipartite graph between user and message is constructed to model homogeneous and heterogeneous interaction relations, which are extracted based on the interaction behaviors.
- We provide a co-clustering algorithm based on non-negative matrix tri-factorization to detect anomalous users and messages simultaneously.
- The homogeneous relations are integrated into co-clustering algorithm as the constraint conditions, which can improve the accuracy of the algorithm.
- An empirical study of our methods on real-world Sina Weibo data, which reveals that the methods are effective for anomaly detection.

The rest of the paper is organized as follows. We describe related works in Section 2. Some anomalous cases and the bipartite graph model are provided in Section 3. We present an algorithm based on non-negative matrix tri-factorization for anomaly detecting in Section 4. Experiment is in Section 5. Section 6 draws conclusions.

2   Related Work

There are many studies on twitter[1, 2], however, some studies found that there are many differences between Sina Weibo and Twitter[3, 4]. The recent researches focus on detecting anomalous users or messages individually[6, 7], but spammers are become more intelligent. In order to evade the Sina Weibo's own anomaly detection system, anomalous user may post some normal messages[8], so some intelligent anomalous users may be missed.

A lot of anomaly detection methods had been proposed, such as machine learning[10] and data mining[11], but they only consider content of message or profile of user, so the accuracy is still low than expected. We will use co-clustering algorithm to analyze user and message simultaneously[9], and consider not only the user attributes and content of message, but also consider the relationships between user and message.

In microblogging, user can generate many interactions between user and message, which facilitate dissemination of information[5]. Bipartite graph is usually used to model the relations between two entities. Some algorithms based on bipartite graph are proposed to detect anomalies[12, 13], but it only considered positive and negative cases. We will consider five user interaction behaviors, including homogeneous and heterogeneous interactions.

3   Anomaly Analysis and Modeling

In this section, through the analysis of anomalous cases, some important attributes have been extracted for anomaly detection. An interaction model between user and message is built to model homogeneous and heterogeneous interactions simultaneously.

3.1   Anomalous Case Analysis

Sina Weibo is one of the most popular social network platforms in China. A variety of operating interfaces were provided to generate content and follow relationship, such as post, retweet, comment, mention, follow et al. A user posts an original message is up to140 characters, called tweet. A user reposts a tweet to his or her followers called retweet. A user comments on a follower's tweet, which is visible to his followers. Any tweet can mentions some users, contains tags, pictures, videos, et al.

Through operating interfaces, a large number of tweets are posted and propagated. However, there are some anomalous messages in Sina Weibo like spammer, advertising, sweepstake and promotion. With the improved of techniques, anomalous users are increasingly intelligent.

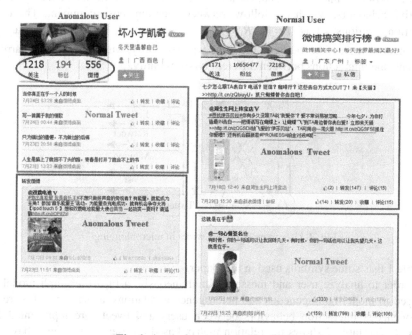

Fig. 1. Anomalous cases in Sina Weibo

Figure 1 shows two user's profile, which are difficult to be detected. Anomalous messages are marked red box, and normal messages are marked blue box.

The left of figure 1 shows an anomalous user. In order to increase the activity and fraudulence, anomalous users may post some normal messages. Another case is shown in the right of figure 1, the normal user often posted normal message, occasionally posted anomalous message.

Traditional algorithms are based on the assumption that messages are posted by anomalous user are anomalous messages, and normal users do not post normal messages. However, the real situation is that anomalous users may post some normal messages, and normal users can also post or retweet anomalous messages. Through the above analysis, the assumption no longer holds.

In order to detect anomaly case in figure 1, the anomaly detection algorithm should not only consider the attribute of users and content of messages, but also consider the interactions between users and messages.

3.2   Features Extraction and Modeling

In Sina Weibo, the user can produce multiple interactions by only one operation, which is different form twitter. For example, while a user retweet a message, he can comment on the message and also mention some users. In order to analyze the user behavior in microblogging, we only consider two types of entities, user and tweet. Figure 2 shows all interaction behaviors between two entities. In this paper, we consider five behaviors, which are follow, retweet, mention, post, comment. The user interaction relations set is $I = \{F, R, M, P, C\}$. In figure 2, the dotted line arrow represents the interactions between the same types of entity, while the solid line arrow represents the interactions between different types of entity.

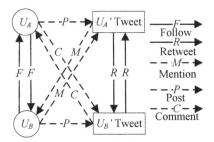

Fig. 2. User interaction behaviors in microblogging

Table 1 lists some symbols used in this paper.

In order to analyze user and message simultaneously, a bipartite graph model is proposed to model homogeneous and heterogeneous relations, as shown in Figure 3. Heterogeneous interaction behaviors between users and tweets are represented by matrix B. The table 2 shows the relation matrix based on user interaction behaviors. Base on matrix B, we proposed a co-clustering algorithm to process tweets and users simultaneously.

Table 1. Symbols

Symbols	Description	Symbols	Descriptions
U	User relations matrix	U^{sim}	User's similar matrix
T	Tweet relations matrix	U^{dis}	User's dissimilar matrix
B	User-Tweet relations matrix	T^{sim}	Tweet's similar matrix
$\tilde{B}$	Constraint relations matrix	T^{dis}	Tweet's dissimilar matrix
F_U	User anomaly matrix	L_U	User distance metric
F_T	Tweet's feature value vector	L_T	Tweets distance metric

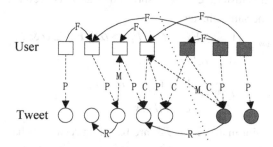

Fig. 3. User-Message interaction model for anomaly detection

The value of matrix B is calculated by Eq. 1. It is difficult to get all interactions between users and messages, so we only consider the types of interactions, and do not consider the times of interactions.

$$B_{i,j} = \begin{cases} 1 & (i \le j \ \& \ behavior\ is\ P\ or\ C) | (i > j \ \& \ behavior\ is\ M) \\ 0 & others \end{cases} \tag{1}$$

According the description in section 3.1, anomalous users and tweets have their own unique characteristics. In this paper, we not only consider user's attribute and the content of message, but also consider interaction relations between user and tweet. For a user U_a, the value of user anomaly matrix F_{U_a} is calculated by Eq. 2, $N_{follower}^{U_a}$ is the number of followers, and $N_{following}^{U_a}$ is the number of followings. For a tweet T_x, the value of tweet's features value vector F_{T_x} is calculated by Eq. 3, $N_{link}^{T_x}, N_{mention}^{T_x}, N_{picture}^{T_x}, N_{hashtag}^{T_x}$ are the number of links, mentions, pictures and hashtags respectively, $|T_x|$ is the length of tweet.

$$F_{U_a} = \frac{N_{follower}^{U_a}}{N_{following}^{U_a}} \tag{2}$$

$$F_{T_x} = 0.5 \times \left(\frac{N_{link}^{T_x} + N_{mention}^{T_x} + N_{picture}^{T_x} + N_{hashtag}^{T_x}}{4} \right) + 0.5 \times \left(1 - \frac{|T_x|}{140} \right) \tag{3}$$

Table 2. Heterogeneous relations based on user behaviors

	User a's Tweet	User b's Tweet
User a	P	C/M
User b	C/M	P

The matrix B is very sparsely, so homogenous interaction relations are used to build constraint matrix. The user constraint matrix U^{sim}, U^{dis} are based on following behaviors. The value of matrix U^{sim}, U^{dis} are calculated by Eq. 4 and Eq. 5. α is the threshold of anomalous user. $U_{a,b}^{sim} = 1$ shows that user U_a and U_b are similarly in the same cluster. $U^{dis} = 1$ shows that user U_a and U_b are dissimilarly, and can't be in the same cluster.

$$U_{a,b}^{sim} = \begin{cases} 1 & (F_{U_a} > \alpha \& F_{U_b} > \alpha)|(F_{U_a} \leq \alpha \& F_{U_b} \leq \alpha) \\ 0 & others \end{cases} \tag{4}$$

$$U_{a,b}^{dis} = \begin{cases} 1 & (F_{U_a} \leq \alpha \& F_{U_b} > \alpha) \\ 0 & others \end{cases} \tag{5}$$

The tweet constraint matrix T^{sim}, T^{dis} are based on retweet behaviors. The value of T^{sim}, T^{dis} are calculated by Eq. 6 and Eq. 7. β is the threshold of anomalous tweet.

$$T_{x,y}^{sim} = \begin{cases} 1 & U_{T_x} follow U_{T_y} \& ((F_{T_y} \geq \beta \& F_{U_{T_x}} \leq \alpha)|(F_{T_y} < \beta \& F_{U_{T_x}} \geq \alpha)) \\ 0 & others \end{cases} \tag{6}$$

$$T_{x,y}^{dis} = \begin{cases} 1 & U_{T_x} not follow U_{T_y} \& (F_{T_y} < \beta \& F_{U_{T_x}} \leq \alpha) \\ 0 & others \end{cases} \tag{7}$$

4   Anomaly Detection Based on Nonnegative Matrix Tri-Factorization

We treat the problem of detecting anomalous users and tweets simultaneously as a co-clustering problem. The nonnegative matrix tri-factorization (NMTF) algorithm and heterogeneous relation metric learning are introduced in this section.

4.1   Heterogeneous Relation Metric Learning

In order to improve the performance of co-clustering algorithm, the homogenous interaction relations are integrated through the distance metric learning.

The distance metric L_U, L_T can be learned from Eq. 8 and Eq. 9, which are the generalized semi-supervised linear discriminate analysis. The constraint matrix U^{sim}, U^{dis}, T^{sim}, T^{dis} are calculated by Eq. 4-7. The detail calculation process can be obtained accordingly[14].

$$Q_U = \min \frac{trace(L_U U^{sim})}{trace(L_U U^{dis})} \tag{8}$$

$$Q_T = \min \frac{trace(L_T T^{sim})}{trace(L_T T^{dis})} \tag{9}$$

In Eq. 8 and Eq. 9, $trace()$ is the trace of matrix.

Through learning the distance metric L_U, L_T, the homogenous interactions are embed into heterogeneous relation matrix B. Through Eq. 10, the original relation matrix B is projected into a new space. The new heterogeneous relation matrix $\tilde{B}$ is provided for co-clustering algorithm.

$$\tilde{B} = \sqrt{L_T} \sqrt{L_U} B \tag{10}$$

4.2 Anomaly Detection Algorithm

After the distance metric learning, we can formulate the task of co-clustering as an optimization problem with nonnegative matrix tri-factorization for $\tilde{B}$. The optimization objective function Q is provided to partition the user and tweet simultaneously.

$$Q = \min_{P_U \geq 0, P_M \geq 0} \left\| \tilde{B} - P_U S P_M \right\|_F^2 \tag{11}$$

Where P_U is the user partition indicator matrix, P_M is the tweet partition indicator matrix. S is the cluster association matrix, which provides the relation between users and tweets. The overall anomaly detection algorithm is shown in **algorithm 1**.

Algorithm1. Anomaly Detection based on Nonnegative Matrix Tri-Factorization (NMTF)

Input: user-tweet matrix: B, U^{sim}, U^{dis}, T^{sim}, T^{dis}

Output: the user and tweet partition indicator matrix: P_U, P_M

1:   Initialize P_U, P_M, S and $K=2$;

2:   Learn the distance metric L_U, L_T based on U^{sim}, U^{dis}, T^{sim}, T^{dis};

3:   Calculate the new relation matrix $\tilde{B}$ based on B, L_U, L_T;

4:       Take Eq. 11 as objective function, iterate update P_U, P_M and S;

5:   $$P_U \leftarrow P_U \frac{S^T P_M^T \tilde{B}}{S^T P_M^T P_M S P_U} \tag{12}$$

6:   $$P_M \leftarrow P_M \frac{\tilde{B} P_U^T S^T}{P_M S P_U P_U^T S^T} \tag{13}$$

7:   $$S \leftarrow S \frac{P_M^T \tilde{B} P_U^T}{P_M^T P_M S P_U P_U^T} \tag{14}$$

8:       Until convergence

9:   Return the partition indicator matrix P_U, P_M

In step2, 3, we take the homogenous interactions as constraint condition, which are embed into heterogeneous relation matrix B by distance metric learning.

In order to obtain the local optimal solution for the objective function Eq. 11, the cluster structure for user and tweet are updated iteratively. In step 5-7, we derive an EM (Expectation Maximization) style approach that iteratively performs the matrix decomposition using a set of multiplicative updating rules.

In the algorithm 1, we set $K=2$, so the returned partition indicator matrix P_U, P_M can easy to distinguish between normal and anomaly.

5   Experiments

In this section, we empirically demonstrate the performance of the proposed method on Sina Weibo data sets.

5.1   Datasets

In experiment, datasets are collected from Sina Weibo. In order to collect the real anomalous users, we purchased 1000 anomalous accounts from Taobao[2]. The partial accounts are detected by Sina Weibo's own anomalous detection system directly, and 778 anomalous accounts are included in our datasets.

We collected 66283 normal users randomly, and the first page's tweets of each user are collected in the experiment. In order to collect interaction behaviors as much as possible, we collected interaction behaviors data separately. The detail of dataset is shown in table 3.

During the preprocessing, we sort the messages of user according to the post time, and extract hashtags, links, pictures and mention user in each message. The structural data are prepared for anomaly detection algorithm.

Table 3. Dataset description

	Users		Tweets	
	Normal	Anomaly	Normal	Anomaly
Number	66283	778	1819568	942325

5.2   Evolution Metrics

In order to evaluate the effectiveness of our anomaly detection method based on non-negative matrix tri-factorization, we use the standard information retrieval metrics viz. precision, recall and F1-score. Detection of a cluster is the proportion of detected positives in cluster that are actually positive. Recall of a cluster is the proportion of the labeled positives in cluster, which are detected positive. In order to explain this further, we use the 'confusion matrix' described in Table 4.

---

[2] http://www.taobao.com

Table 4. Confusion matrix for cluster

	Anomaly Detection	Normal Detection
Anomaly Label	*tp*	*fn*
Normal Label	*fp*	*tn*

Each entry in the table indicates the number of elementsin a cluster and the way they were detected. By using thisconfusion matrix, the precision, recall and F1-score are respectively defined as:

$$Precision = \frac{tp}{tp + fp} \tag{15}$$

$$Recall = \frac{tp}{tp + fn} \tag{16}$$

$$F1 - Score = \frac{2 \times Precision \times Recall}{Precision + Recall} \tag{17}$$

5.3   Results and Discussion

For anomalous users detecting, if the number of messages is very big, the processing performance of the algorithm, vice versa, and the accuracy of the algorithm will be affected. We need to verify how many messages needed for anomalous users detecting. For every user, we extract the message on the first page, which is ordered by time. In figure 4, the result of varying the number of messages shows that the F1-measure is higher when we extract ten messages.

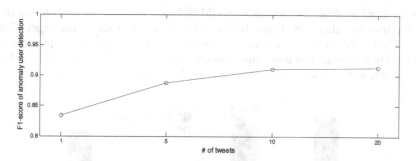

Fig. 4. The result of varying the number of messages

To empirically study the effectiveness of our NMTF-based method, we compared the accuracy of NMTF-based method, NMF-based method and SVM-based method. NMF-based method is nonnegative matrix factorization without integrated constraint conditions. SVM-based method is a classical classification method to detect spammers. We use LibSVM[15] as the baseline classifier method. For anomaly detection, we use user and tweet as feature vector, and train two SVM model.

In comparative experiments, each user extracted ten messages, the average results of anomalous users and messages detecting with algorithm 1 are shown in figure 5. The experimental results show that NMTF-based method has the higher accuracy than NMF-based method and SVM-based method in both anomalous users and messages detection. This is mainly due to both homogeneous and heterogeneous interactions are considered in the NMTF-based method. SVM is a classic classification method, however, the accuracy of SVM-based method is sensitive to the characteristics of the data.

In the anomalous user detection experiment, although the characteristics of anomalous users are more obviously, the accuracy of anomalous users detection are lower than the accuracy of normal user detection. Through the analysis, the possible reason is anomalous users post a lot of normal messages, but fewer anomalous messages.

In the anomalous message detection experiment, the accuracy of anomalous and normal messages detection are not high. Since a lot of messages have few features, it is hard to judge if an anomalous message is posted by a normal user.

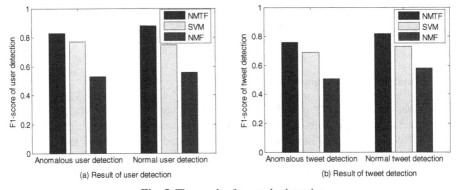

Fig. 5. The result of anomaly detection

For in-depth analysis the accuracy of NMTF-based method, we analyze the message and user in detail. The figure 6 shows four characteristics of messages. Compared with normal messages, Anomalous messages contain more links, pictures and hashtags, but the number of mentions is roughly the same. Anomalous messages have more obvious features, so the accuracy of anomalous detection is very high.

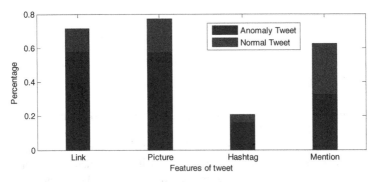

Fig. 6. The statistical characteristics of messages

The figure 7 shows the ratio of users' follower number and following number. For the normal user, the user's follower number is more than following number, which is shown in the figure 7. The most anomalous users' follower number is less than following number, but there are some exceptions, which are shown in the figure 7 of the above. The user named '@Love-Constellation Shopping Fashion' and '@Miko Sweater Channel' are detected as anomalous users marked in the figure 7. The user's follower number is more than following number, this feature is the same as the normal user, so the traditional method can't detect such users. Because the user usually post normal messages, occasionally post some promotional messages, so it is difficult to detect.

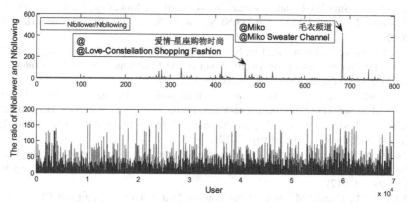

Fig. 7. The ratio of users' follower number and following number

6   Conclusions

This paper focuses on the analysis of homogeneous and heterogeneous interaction realations of user and message in Sina Weibo, which are modeled as a Bipartite graph. We proposed a co-clustering algorithm based on nonnegative matrix tri-factorization to detect anomalous users and messages simultaneously. The homogeneous interactions were integrated into co-clustering algorithm and improved the accuracy of the algorithm. The experiment results show that the accuracy of our method is very high.

In the future work, we will extend our method to online mode, and extract interactions to build a dynamic bipartite graph for online anomaly detection based on real-time message steams.

Acknowledgments. The work is supported by the Natioal Natural Science Foundation of China(61170242), 863 High Technology Program(2012AA012802), and Fundamental Research Funds for the Central Universities(HEUCF100611).

References

1. Kwak, H., Lee, C., Park, H., Moon, S.: What is Twitter, a Social Network or a News Media? In: Proceedings of the 19th International Conference on World Wide Web, pp. 591–600. ACM, New York (2010)
2. Wu, S., Hofman, J.M., Mason, W.A., Watts, D.J.: Who Says What to Whom on Twitter. In: Proceedings of the 20th International Conference on World Wide Web, pp. 705–714. ACM, New York (2011)
3. Yu, L., Asur, S., Huberman, B.A.: What Trends in Chinese Social Media. In: Proceedings of the 5th SNA-KDD Workshop (2011)
4. Gao, Q., Abel, F., Houben, G.-J., Yu, Y.: A Comparative Study of Users' Microblogging Behavior on Sina Weibo and Twitter. In: Masthoff, J., Mobasher, B., Desmarais, M.C., Nkambou, R. (eds.) UMAP 2012. LNCS, vol. 7379, pp. 88–101. Springer, Heidelberg (2012)
5. Zhu, Y., Wang, X., Zhong, E., Liu, N.N., Yang, Q.: Discovering Spammers in Social Networks. In: Proceedings of the Twenty-Sixth AAAI Conference on Artificial Intelligence (AAAI 2012), Toronto, Ontario, Canada, pp. 172–177 (2012)
6. Bosma, M., Meij, E., Weerkamp, W.: A Framework for Unsupervised Spam Detection in Social Networking Sites. In: Baeza-Yates, R., de Vries, A.P., Zaragoza, H., Cambazoglu, B.B., Murdock, V., Lempel, R., Silvestri, F. (eds.) ECIR 2012. LNCS, vol. 7224, pp. 364–375. Springer, Heidelberg (2012)
7. Chu, Z., Widjaja, I., Wang, H.: Detecting Social Spam Campaigns on Twitter. In: Bao, F., Samarati, P., Zhou, J. (eds.) ACNS 2012. LNCS, vol. 7341, pp. 455–472. Springer, Heidelberg (2012)
8. McCord, M., Chuah, M.: Spam Detection on Twitter Using Traditional Classifiers. In: Alcaraz Calero, J.M., Yang, L.T., Mármol, F.G., García Villalba, L.J., Li, A.X., Wang, Y. (eds.) ATC 2011. LNCS, vol. 6906, pp. 175–186. Springer, Heidelberg (2011)
9. Juan, M.R., Lourdes, A.: Detecting malicious tweets in trending topics using a statistical analysis of language. Expert Systems with Applications 40(8), 2992–3000 (2013)
10. Chen, Y., Wang, L., Dong, M.: Non-negative Matrix Factorization for Semi Supervised Heterogeneous Data Coclustering. IEEE Transactions on Knowledge and Data Engineering 22(10), 1459–1474 (2010)
11. Jiang, J., Wilson, C., Wang, X., Huang, P., Sha, W.: Understanding Latent Interactions in Online Social Networks. In: Proceedings of the 10th ACM SIGCOMM Conference on Internet Measurement, pp. 369–382. ACM, New York (2010)
12. Dai, H., Zhu, F., Lim, E., Pang, H.: Detecting Anomalies in Bipartite Graphs with Mutual Dependency Principles. In: IEEE 12th International Conference on Data Mining (ICDM 2012) (2012)
13. Sun, J., Qu, H., Chakrabarti, D., Faloutsos, C.: Neighborhood Formation and Anomaly Detection in Bipartite Graphs. In: IEEE 5th International Conference on Data Mining (ICDM 2005) (2005)
14. Xing, E.P., Ng, A.Y., Jordan, M.I., Russell, S.: Distance Metric Learning, with Application to Clusering with Side-information. In: Proceeding of the 16th Neural Information Processing Systems, Vancouver, pp. 505–512 (2002)
15. Chang, C.-C., Lin, C.-J.: LIBSVM: A library for support vector machines. ACM Transactions on Intelligent Systems and Technology 2(3), 27:1–27:27 (2011)

Expanding Native Training Data for Implicit Discourse Relation Classification

Yu Hong, Shanshan Zhu, Weirong Yan, Jianmin Yao,
Qiaoming Zhu, and Guodong Zhou

Key Laboratory of Natural Language Processing of Jiangsu Province
School of Computer Since and Technology, Soochow University,
No.1 Shizi Street, Suzhou City, Jiangsu Province, China
tianxianer@gmail.com

Abstract. Linguistically informed features are provably useful in classifying implicit discourse relations among adjacent text spans. However the state of the art methods in this area suffer from either sparse natively implicit relation corpus or counter-intuitive artificially implicit one, and consequently either insufficient or distorted training in automatically learning discriminative features. To overcome the problem, this paper proposes a semantic frame based vector model towards unsupervised acquisition of semantically and relationally parallel data, aiming to enlarge natively implicit relation corpus so as to optimize the training effect. Experiments on PDTB 2.0 show the usage of the acquired parallel corpus gives statistically significant improvements over that of the prototypical corpus.

Keywords: Implicit discourse relation, Semantic frame, Parallel corpus, Discourse analysis.

1   Introduction

One focus in discourse analysis and modeling is to automatically identify and label the semantic relations between text spans such as causal, temporal, comparable ones, etc. See the causal relation in (1).

(1) He falls down (**because**) he steps onto a ball.

As widely admitted in literature, these relations serve as the basic types of discourse relationships, benefiting discourse generation (Mann and Thompson, 1988; Hobbs, 1990; Lascarides and Asher, 1993; Knott and Sanders, 1998; Webber, 2004; Prasad et al., 2005). Accordingly a discourse relation recognizer is able to promote downstream natural language processing such as summarization (Marcu, 2000b), sentence compression (Sporleder and Lapata, 2005), question answering (Verberne et al., 2007), text ordering (Lin et al., 2011) and clustering (Zahri et al., 2012). For example, recognizing causal relations enable answering "why" questions.

Discourse relation can be explicitly signaled by intuitive markers (see (1)), commonly known as the explicit discourse relation (Marcu et al., 2002; Pitler et al., 2008).

H.-Y. Huang et al. (Eds.): SMP 2014, CCIS 489, pp. 67–75, 2014.
© Springer-Verlag Berlin Heidelberg 2014

The optional markers consist of adverbials (e.g., therefore), conjunctions (e.g., because) or alternative lexicalizations (abbr., AltLex, e.g., result in (Prasad et al., 2010)). Oppositely, the absence of such markers between interrelated text spans gives the birth to the implicit discourse relation (Pitler et al., 2008). See the implicit causal relation between the spans in (2). Our investigation concentrates on the task of implicit discourse relation recognition.

(2) Span 1: A tsunami hit Phuket Island.
        Span 2: The coastal highways were wracked.

As typically represented by lexical and syntactic information, linguistically informed features and the combination of these features are practically useful in supervised or semi-supervised implicit discourse relation classification (Marcu and Echihabi, 2002; Pettibone and PonBarry, 2003; Soricut, 2003; Saito et al., 2006; Wellner et al., 2006; Pitler et al., 2008; Pitler et al., 2009; Lin et al., 2009; Wang et al., 2010; Park and Cardie, 2012; Biran and McKeown, 2013). However the classifiers suffered from the shortage of manually annotated implicit relation examples and consequently insufficient training effect.

Towards the resolution of sparse training examples, earlier studies firstly technically selected pairwise spans containing certain explicit markers from raw corpus, then removed the markers from the spans, and so as to automatically generate the simulated implicit relation corpus (Marcu and Echihabi, 2002; Blair-Goldensohn et al., 2007; Sporleder and Lascarides, 2008). Although better informed in use of such corpus in training, the classifiers were illustrated to not perform as well as expected because most of the artificially implicit relation examples are counter-intuitive and easily misguide the classifier learning procedure (Lin et al., 2009).

With the release of the second version of Penn Discourse Treebank (abbr., PDTB 2, Prasad et al., 2008), which provides 22, 141 manually annotated implicit relation examples, recent studies addressed this area of work on a cleaner and genuine corpus. However, the corpus isn't quite large enough to solve the sparsity issue, especially when used in the acquisition of discriminative lexical features (Biran and McKeown, 2013). Worse yet, Wang et al (2012) illustrated that only those typical relation data in PDTB benefit the relation classification. That is, the available naively implicit relation corpus reduces further.

(3a) Span 1: *Some have raised their cash positions to record levels.*
        Span 2: (Suppositional Marker = because) *High cash positions help buffer a fund when the market falls.*

<PDTB.ID=WSJ0983.47&48>
<Implicit Relation=Causality>

(3b) Span 1: *The amount of cash that some funds have now is stepped up to the top position.*
        Span 2: *Large monetary holdings help ease the impact of the declining market.*

<SRPD >
<Implicit Relation=Causality>

To overcome the problem, we introduce the semantically and relationally parallel data (SRPD) mined from external resources into PDTB corpus, by which to enlarge the available training data for training implicit relation classifiers. A SRPD example is a pair of text spans which have the same semantics and the same type of discourse relation as that of the original PDTB example. See (3a) and (3b) which respectively list a PDTB example and the SRPD to it.

As a SRPD example normally has different linguistic phenomenon from the original, such as distinct lexical pragmatics and syntactic rules in (3b), adding such examples to PDTB is extremely useful to incorporate new instantiated samples of the discriminative linguistic features into the training data and consequently enhance the learning effect of the relation classifiers. For example, other than the word pairs originally provided by PDTB such as the pair "raise" and "buffer" in (3a), SRPD is able to merge new semantically related word pairs into the training data, such as "step up" and "ease" in (3b), and thus prevent the co-occurred word pair based classification (Marcu and Echihabi, 2002; Lin et al., 2009; Pitler et al., 2009; Biran and McKeown, 2013) from missing finding many relationally discriminative word pairs in the training data.

Towards the acquisition of SRPD, we propose a semantic frame based vector model to support the similarity calculation among text spans at a semantic level. The model describes a text span using a vector of lexical semantic frames. Each frame in use is quoted from FrameNet[1], where a frame serves as a word sense and characterizes a cluster of lexical units that have the same basic concept and act as the consistent semantic role, such as the frame "Death" in representing the units "starve", "asphyxiate", "perish", etc. The usage of such a model benefits the measurement of semantic similarity, and accordingly promotes retrieving the SRPD having the same semantics with PDTB examples even if the SRPD consist of words of different morphologies from that in the PDTB ones. For example, although (3b) is very different in the use of words from (3a), it is still able to be recalled in our approach because the both share the same frame vector as below:

Semantic Frame (sf) Vector{
sf1→<Cause_change_of_position_on_a_scale>
units→ *raised* (in 3a), *stepped up* (in 3b)
sf2→<Rank>
units→ *level* (in 3a), *top* (in 3b)
sf3→<Change_postition_on_a_scale>
units→ *falls* (in 3a), *declining* (in 3b)...}

On the basis, we use Hong et al (2012)'s two-pronged similarity calculation approach to ensure the parallel relations of SRPD to PDTB examples. Experiments show that the usage of SRPD gives statistically significant improvements over that of the prototypical training corpus.

---

[1] https://framenet.icsi.berkeley.edu/fndrupal/home

2   FrameNet

FrameNet is a machine-readable lexical database of English, based on the theory of frame semantics (Fillmore, 1976; Fillmore and Baker, 2001, 2003 and 2012), indexes more than 10,000 semantic frames. Each frame consists of definition (Def.), frame elements (Eles.), role type (Rol.), Successors (Sucs.) and the words or phrases that evoke the frame (also called lexical units) (LUs.). Listed in Table 1 are the components of the semantic frame "Change_position_on_a_scale".

FrameNet gives an index to alphabetical lists of the name of the lexical units (LUs). Each LU is followed by the part of speech, status, and the frames it evokes. In our approach, we use the index to retrieve the frames for the words and/or phrases (viz., LUs) in text spans.

Table 1. An example of Semantic Frame

Frame: "Change_position_on_a_scale"	
Def.	Indicating the change of an item's position on a scale from a starting point to an end point.
Eles.	Attribute, Initial_value, Final_value, Value-range, Initial_state, Final_state, Difference
Rol.	Manner
Sucs.	Change_of_temperature, Proliferating_in_num
LUs.	*decline* (n), *decline* (v), *decrease* (n), *decrease* (v), *depressed* (a), *depression* (n), *drop* (v), *fall* (n&v), *dip* (n), *elevated* (a), *elevation* (n), *accelerated* (a), *advance* (v), *fluctuate* (v), …

In the case that an ambiguous LU evokes more than one frame, we use the following rules (Ruls.) to determine the appropriate frame in a specific text span:

- **Rul. 1:** The frame necessarily subjects to the part of speech the LU occupies in the span.
- **Rul. 2:** If available, the frame subjects to the phrase structure the LU sits in.
- **Rul. 3:** The frame needs to have the frame elements (Eles.) most similar to the frames of the words co-occurred with the LU in the text span.

Listed in Table 2 are the frames evoked by the LU "fall" along with the respective rules available in the frame disambiguation.

Table 2. An example of the rule based frame disambiguation

LU.	POS.	Frame	Rul.
fall	n.	Change_position_on_a_scale	1
fall	v.	Change_position_on_a_scale	1&3
fall	v.	Motion_directional	1&3
fall	v.	Conquering	1&3
fall (on)	VP.	Becoming_aware	2
fall (upon)	VP.	Attack	2

We additionally use Das and Smith (2011)'s graph-based semi-supervised learning method to assign semantic frames to LUs out of vocabulary.

3 Methodology

We use PDTB implicit relation data as seeds to search SRPD in external linguistic resource, such as the third version of English Gigaword[2] used in our experiments. Thereafter we add the acquired SRPD back to PDTB to expand the training data. On the basis, we utilize a SVM classifier well trained on the data to classify discourse relations.

Towards the acquisition of SRPD, we use the proposed semantic frame based vector model to describe text spans regardless of whether the spans are derived from the external resource (e.g., Gigaword) or PDTB. And we follow Hong et al (2012)'s two-pronged similarity calculation between pairs of text spans to ensure the semantic and relational parallelism. Given a pair of adjacent Gigaword spans p_x, $p_x=\{x_{pre}, x_{pos}\}$ and a pair of PDTB spans p_y, $p_y=\{y_{pre}, y_{pos}\}$, where x_{pre} is a prepositive span in p_x in sentence order while x_{pos} is postpositive, in the same way with y_{pre} and y_{pos}, we determine p_x as a qualified SRPD when it has a high two-pronged similarity ($S.$) to p_y, calculated as below:

$$S. = \frac{Sim_{pre}(x_{pre}, y_{pre}) + Sim_{pos}(x_{pos}, y_{pos})}{2 \cdot e^{|Sim_{pre} - Sim_{pos}|}} \tag{1}$$

where, the exponential function is to ensure nonzero denominator and the number 2 normalizes the whole similarity, and either Sim_{pre} or Sim_{pos} is calculated using VSM based Cosine metric.

For each seeded PDTB example, we select n pairs of spans that have the highest two-pronged similarity to it as the qualified SRPD. Thereafter we use the qualified SRPD to expand the prototypical PDTB training data.

We build the SVM classifier using several superior and steady features in previous work, including **Word Pairs** (Marcu and Echihabi, 2002), **Verbs** (Piter et al., 2009), **Production** Rules & **Dependency** (Lin et al., 2009), **Syntax** (Wang et al., 2010) and **Similarity** (Biran and McKeown, 2013). In experiments we concentrate on verifying the effect of using SRPD to enhance the features in classifying implicit discourse relations.

4 Experiments

We use implicit discourse relation examples in sections No.00-20 of the 2nd version of PDTB as the prototypical training data, including 13,815 examples in total, sections No. 21&22 as development data, including 1,046 examples, and section No. 23&24 as test data, including 1,192 examples. Only the top-level relation senses are considered in our experiments such as Expansion, Contingency, Comparison and Temporal. Be advised that the relation examples of EntRel and NoRel are not included in any of the above training, development or test data (see the data distribution in PDTB v2 Annotation Manual (Prasad et al., 2007)). Additionally, we randomly select one million pairs of adjacent text spans from all news stories of NYT in the 3rd version of English Gigaword and use them as the designated external resources for SRPD mining.

---

[2] https://catalog.ldc.upenn.edu/LDC2009T28

To check the efficacy of the different feature classes, we use LibSVM toolkit to build individual classifiers on all features within a single feature class. The classifiers are finally evaluated using accuracy (Acc.).

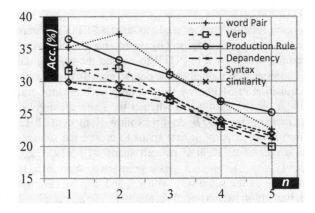

Fig. 1. Accuracies of all individual classifiers on development data when trained on different sets of SRPD acquired with n ranges from 1 to 5

Firstly, we use the development data to confirm the optimal parameter n which is the presupposed number of the qualified SRPD to each seeded PDTB example. We use the whole training data as seeds and obtain 5 different sets of SRPD from the designated external resources when changing n from 1 to 5. Thereafter we train the classifiers for the quaternary relations respectively on the 5 SRPD sets and debug them on the development data. Figure 1 shows the four-way accuracies of the classifiers achieved at different n. According to the results, we set n equal to 1 on which most classifiers are superior and the least computational expense is needed to acquire qualified SRPD.

Table 3. Accuracies of the classifiers achieved on test data after going through different training

	Verb	Word Pair	Similarity
Proto.	34.1%	35.6%	34.3%
Base.	8.30%	10.9%	9.50%
Base.+Proto.	19.6%	22.6%	21.6%
SRPD	30.2%	31.6%	30.1%
SRPD+Proto.	**40.3%**	**43.5%**	**36.1%**
	Production	Syntax	Dependency
Proto.	38.3%	33.3%	32.9%
Base.	12.8%	11.7%	8.39%
Base.+Proto.	26.2%	19.2%	18.5%
SRPD	33.9%	28.9%	28.5%
SRPD+Proto.	**43.2%**	**36.9%**	**35.5%**

Secondly, we add the optimal SRPD set ($n=1$) back to the prototypical training data. Using the expanded training data, we train the classifiers for the quaternary relations (four-way). For the sake of contrastive analysis, for each PDTB example in the prototypical training data, we randomly select a pair of adjacent text spans from the designated external resources to create the pseudo-SRPD. We use all pseudo-SRPD as baseline (Base.) because they are the most unreliable training data. Such SRPD are also added to the prototypical training data to jointly train the classifiers. Besides we train the classifiers on the prototypical training data (Proto.) as normal. Finally, we evaluate the classifiers on the test data.

Table 3 shows the accuracies of the classifiers (features) on test data. The results illustrate that adding SRPD to the prototypical training data enhance all features in discriminating implicit relations. We suggest that the improvements derive from the additionally acquired qualified linguistic knowledge on textual implicit relations. By contrast, involving raw external resources into training data negatively influences classifier learning, such as performance reduction caused by the randomly selected examples in baseline.

5   Conclusion

In this paper we illustrate the feasibility of the semantic frame vector model in acquiring qualified naively implicit relation data. Additionally we prove such data can be used to expand training data and further optimize current linguistic feature based discourse relation classification.

Through this study, we try to raise awareness of the necessity to enrich linguistic knowledge on discourse relations. A sufficient knowledge base is helpful to energize the existing features in distinguishing among relations. It is extremely important when few features are available. For example, syntax and dependency aren't available in analyzing relations among event mentions. That is because interrelated events aren't always described in adjacent text spans but disjunctively or even in different documents (cross-document relations) while syntax and dependency are only available on sequential sentences. In this case, only few features can be used in relation classification such as Word Pairs. Thus it is necessary to enlarge linguistic knowledge on such features for better learning, by which to compensate for the loss caused by withdrawing some once-mighty but unavailable features.

Acknowledgments. This research is supported by the National Natural ScienceFoundation of China (No. 61373097, 61272259, 61272260, 90920004), the Special fund project of the Ministry of Education Doctoral Program (2009321110006, 20103201110021), the Natural Science Foundation of Jiangsu Province (No.BK2011282), the Major Project of College Natural Science Foundation of Jiangsu Province (No. 11KIJ520003) and the Natural Science Foundation of Jiangsu Province, Suzhou City (SYG201030).

References

1. Mann, W.C., Thompson, S.A.: Rhetorical Structure Theory: Toward a functional theory of text organization. Text 8(3), 243–281 (1988)
2. Hobbs, J.R.: Literature and cognition. CSLI Lecture Notes, vol. 21. CSLI Publications (1990)

3. Lascarides, A., Asher, N.: Temporal interpretation, discourse relations and commonsense entailment. Linguistics and Philosophy 16(5), 437–493 (1993)
4. Knott, A., Sanders, T.: The classification of coherence relations and their linguistic markers: An exploration of two languages. Journal of Pragmatics 30(2), 135–175 (1998)
5. Webber, B.: D-LTAG: Extending lexicalized TAG to discourse. Cognitive Science 28(5), 751–779 (2004)
6. Prasad, R., Joshi, A., Dinesh, N., Lee, A., Miltsakaki, E., Webber, B.: The Penn Discourse TreeBank as a Resource for Natural Language Generation. In: Proceedings of the Corpus Linguistics Workshop on Using Corpora for Natural Language Generation, Birmingham, U.K., pp. 25–32 (2005)
7. Marcu, D.: The Theory and Practice of Discourse Parsing and Summarization. MIT Press, Cambridge (2000b)
8. Sporleder, C., Lapata, M.: Discourse Chunking and its Application to Sentence Compression. In: Proceedings of the Conference on Human Language Technology and Empirical Methods in Natural Language Processing (EMNLP 2005), Vancouver, British Columbia, Canada, pp. 257–264 (2005)
9. Verberne, S., Boves, L., Oostdijk, N., Coppen, P.: Evaluating Discourse-based Answer Extraction for Why-question Answering. In: Proceedings of the 30th Annual International ACM SIGIR Conference on Research and Development in Information Retrieval (SIGIR 2007), Amsterdam, The Netherlands, pp. 735–736 (2007)
10. Lin, Z., Tou Ng, H., Kan, M.: Automatically Evaluating Text Coherence Using Discourse Relations. In: Proceedings of the 49th Annual Meeting of the Association for Computational Linguistics: Human Language Technologies (ACL-HLT 2011), Portland, Organ, USA, pp. 997–1006 (2011)
11. Zahri, N.A.H.: Exploiting Discourse Relations between Sentences for Text Clustering. In: Proceedings of the Workshop on Advances in Discourse Analysis and its Computational Aspects (ADACA, COLING 2012), Mumbai, India, pp. 17–31 (2012)
12. Marcu, D., Echihabi, A.: An Unsupervised Approach to Recognizing Discourse Relations. In: Proceedings of the 40th Annual Meeting of the Association for Computational Linguistics (ACL 2002), Philadelphia, USA, pp. 368–375 (2002)
13. Pitler, E., Raghupathy, M., Mehta, H., Nenkova, A., Lee, A., Joshi, A.: Easily Identifiable Discourse Relations. In: Proceedings of the 22nd International Conference of Computational Linguistics (COLING 2008), Manchester, UK, pp. 87–90 (2008)
14. Prasad, R., Joshi, A., Webber, B.: Realization of Discourse Relations by Other Means: Alternative Lexicalizations. In: Proceedings of the 24th International Conference of Computational Linguistics (COLING 2010), Beijing, China, pp. 1023–1031 (2010)
15. Pettibone, J., PonBarry, H.: A Maximum Entropy Approach to Recognizing Discourse Relations in Spoken Language. Working Paper. The Stanford Natural Language Processing Group (June 6, 2003)
16. Soricut, R., Marcu, D.: Sentence Level Discourse Parsing Using Syntactical and Lexical Information. In: Proceedings of Human Language Technology and North American Association for Computational Linguistics (HLT-NAACL 2003), Edmonton, Canada, pp. 149–156 (2003)
17. Saito, M., Yamamoto, K., Sekine, S.: Using Phrasal Patterns to Identify Discourse Relations. In: Proceedings of Human Language Technology and North American Association for Computational Linguistics (HLT-NAACL 2006), New York, USA, pp. 133–136 (2006)
18. Wellner, B., Pustejovsky, J., Havasi, C., Rumshisky, A., Sauri, R.: Classification of Discourse Coherence Relations: An Exploratory Study Using Multiple Knowledge Sources. In: Proceedings of the 7th SIGDIAL Workshop on Discourse and Dialogue, Sydney, Australia, pp. 117–125 (2006)

19. Pitler, E., Louis, A., Nenkova, A.: Automatic Sense Prediction for Implicit Discourse Relations in Text. In: Proceedings of the Joint Conference of the 47th Annual Meeting of the Association for Computational Linguistics and the 4th International Joint Conference on Natural Language Processing of the Asian Federation of Natural Language Processing (ACL-IJCNLP 2009), Suntec, Singapore, pp. 683–691 (2009)

20. Lin, Z., Kan, M., Ng, H.T.: Recognizing Implicit Discourse Relations in the Penn Discourse Treebank. In: Proceedings of the 2009 Conference on Empirical Methods in Natural Language Processing (EMNLP 2009), Singapore, pp. 343–351 (2009)

21. Wang, W., Su, J., Tan, C.L.: Kernel Based Discourse Relation Recognition with Temporal Ordering Information. In: Proceedings of the 48th Annual Meeting of the Association for Computational Linguistics (ACL 2002), Uppsala, Sweden, pp. 710–719 (2010)

22. Park, J., Cardie, C.: Improving Implicit Discourse Relation Recognition Through Feature Set Optimization. In: Proceedings of the 13th Annual Meeting of the Special Interest Group on Discourse and Dialogue (SIGDIAL 2012), Seoul, South Korea, pp. 108–112 (2012)

23. Biran, O., McKeown, K.: Proceedings of the 51st Annual Meeting of the Association for Computational Linguistics (ACL 2013), Sofia, Bulgaria, pp. 69–73 (2013)

24. Blair-Goldensohn, A., McKeown, K., Rambow, O.: Building and Refining Rhetorical-Semantic Relation Models. In: Proceedings of Human Language Technology and North American Association for Computational Linguistics (HLT-NAACL 2007), Rochester, New York, USA, pp. 428–435 (2007)

25. Sporleder, C., Lascarides, A.: Using Automatically Labeled Examples to Classify Rhetorical Relations: An Assessment. Natural Language Engineering 14(03), 369–416 (2008)

26. Prasad, R., Dinesh, N., Lee, A., Miltsakaki, E., Robaldo, L., Joshi, A., Webber, B.: The Penn Discourse TreeBank 2.0. In: Proceedings of the 6th International Conference on Language Resources and Evaluation (LREC 2008), Marrakech, Morocco, pp. 2961–2968 (2008)

27. Wang, X., Li, S., Li, J., Li, W.: Implicit Discourse Relation Recognition by Selecting Typical Training Examples. In: Proceedings of the 26th International Conference of Computational Linguistics (COLING 2012), Mumbai, India, pp. 2757–2772 (2012)

28. Hong, Y., Zhou, X., Che, T., Yao, J., Zhou, G., Zhu, Q.: Cross-Argument Inference for Implicit Discourse Relation Recognition. In: Proceedings of the 21st ACM International Conference on Information and Knowledge (CIKM 2012), Maui, HI, USA, pp. 295–304 (2012)

29. Fillmore, C.J.: Frame Semantics and Nature of Language. In: Annals of the New York Academy of Scienes: Conference on the Origin and Development of Language and Speech, vol. (280), pp. 20–32 (1976)

30. Fillmore, C.J., Baker, C.F.: Frame Semantics for Text Understanding. In: Proceedings of WordNet and Other Lexical Resources Workshop (NAACL 2001), Pittsburgh, USA (2001)

31. Fillmore, C.J., Baker, C.F.: The Structure of the FrameNet Database. International Journal of Lexicography 16(3), 281–296 (2003)

32. Fillmore, C.J.: FrameNet, Current Collaborations and Future Goals. Language Resources and Evaluation (46), 269–286 (2012)

33. Das, D., Smith, N.A.: Semi-Supervised Frame-Semantic Parsing for Unknown Predicates. In: Proceedings of the 49th Annual Meeting of the Association for Computational Linguistics: Human Language Technologies (ACL-HLT 2011), Portland, Organ, USA, pp. 1435–1444 (2011)

34. Prasad, R., Miltsakaki, E., Dinesh, N., Lee, A., Joshi, A.: The Penn Discourse Treebank 2.0 Annotation Manual. Technical Report IRCS-08-01, Institute for Research in Cognitive Science, University of Pennsylvania (2007),
http://www.seas.upenn.edu/~pdtb/PDTBAPI/
pdtb-annotation-manual.pdf

Microblog Sentiment Analysis with Emoticon Space Model*

Fei Jiang, Yiqun Liu, Huanbo Luan, Min Zhang, and Shaoping Ma

State Key Laboratory of Intelligent Technology and Systems,
Tsinghua National Laboratory for Information Science and Technology,
Department of Computer Science and Technology,
Tsinghua University, Beijing 100084, China
{f91.jiang,luanhuanbo}@gmail.com, {yiqunliu,z-m,msp}@tsinghua.edu.cn

Abstract. Emoticons have been widely employed to express different types of moods, emotions and feelings in microblog environments. They are therefore regarded as one of the most important signals for microblog sentiment analysis. Most existing works use several emoticons that convey clear emotional meanings as noisy sentiment labels or similar sentiment indicators. However, in practical microblog environments, tens or even hundreds of emoticons are frequently adopted and all emoticons have their own unique emotional signals. Besides, a considerable number of emoticons do not have clear emotional meanings. An improved sentiment analysis model should not overlook these phenomena. Instead of manually assigning sentiment labels to several emoticons that convey relatively clear meanings, we propose the emoticon space model (ESM) that leverages more emoticons to construct word representations from a massive amount unlabeled data. By projecting words and microblog posts into an emoticon space, the proposed model helps identify subjectivity, polarity and emotion in microblog environments. The experimental results for a public microblog benchmark corpus (NLP&CC 2013) indicate that the ESM effectively leverages emoticon signals and outperforms previous state-of-the-art strategies and benchmark best runs.

Keywords: Microblog Sentiment Analysis, Emoticon Space, Polarity Classification, Subjectivity Classification, Emotion Classification.

1 Introduction

Microblogs, such as Twitter and Sina Weibo, are a popular social media in which millions of people express their feelings, emotions, and attitudes. Because a large number of microblog posts are generated every day, the mining of sentiments from this data source assists in the performance of research on various topics, such as analyzing the reputation of brands [12], predicting the stock market

---

* This work was supported by Tsinghua-Samsung Joint Lab and Natural Science Foundation (61472206, 61073071, 61303075) of China.

H.-Y. Huang et al. (Eds.): SMP 2014, CCIS 489, pp. 76–87, 2014.

[5] and detecting abnormal events [22]. Therefore, an improvement in the performance of sentiment analysis tasks in microblog environments is crucial. Microblog sentiment analysis has been a hot research area in recent years, and several important issues have been studied, such as identifying whether a post is subjective or objective (subjectivity classification) [13,15], identifying whether a post is positive or negative (polarity classification) [13,15], and recognizing the emotion in a particular post (emotion classification) [22].

Supervised machine learning techniques have been widely adopted to microblog sentiment analysis and have proven to be effective [3]. Various features, such as the sentiment lexicons, the part-of-speech tags, and microblogging features have been exploited to reinforce the classifiers [14,1]. However, manually labeling sufficient training posts is extremely labor intensive because of the large vocabulary adopted by microblog users. Fortunately, in microblog environments, various emoticons are frequently adopted and are usually posted along with emotional words. In our investigation, there are more than a thousand different emoticons in Sina Weibo, and microblog posts containing emoticons take up a proportion of over 25% in a dataset of about 20 million posts randomly collected from March 2012 to December 2012 (Dataset 1). Moreover, graphical emoticons, which are more vivid compared with those composed of punctuation marks, have been introduced in many microblog platforms. Thus, emoticons can serve as an effective source of emotional signals, making it possible to perform sentiment classification tasks without or with a small amount of manually labeled posts.

A small portion of emoticons has very clear emotional meanings, such as 😀 and 😁 for positive sentiment, ● and 😩 for negative sentiment. Many existing works use these emoticons as noisy sentiment labels of microblog posts to train classifiers [20,10,22] and to avoid the expensive cost of manually labeling data. However, the performance of these research efforts is affected by several problems. First, as contexts and user habits vary, the sentiment of the words in such posts may not be the same as the emoticons. Thus, noises are introduced. Second, every emoticon has a unique emotional meaning in microblog environments. For example, 😀 and 😁 both express an emotion of happiness, but words that co-occur with the latter tend to be more naughty. Such information can be not utilized by equally treating them as the noisy positive label. Moreover, some emoticons, such as 🐶 and 😼 do not have clear emotional meanings and it is hard to assign sentiment labels to the emoticons. They form a considerable portion of the employed emoticons and posts that contain these emoticons have a proportion of over 7% in Dataset 1. These emoticons may also help identify sentiment. In our investigation, the emotion of emoticon 🤖 (Ultraman) is uncertain, but posts containing this emoticon are unlikely to have an emotion of anger.

For the first problem, a state-of-the-art method has been proposed to combine noisy data and manually labeled data to effectively train classifiers [15]. However, no previous models have been proposed to fully exploit the potential of emoticons, by either differently treating each emoticon or integrating those without clear emotional meanings.

To better utilize the signals contained in emoticons and to improve the performance of microblog sentiment classification tasks, we propose a semi-supervised model, called the emoticon space model (ESM). In this model, we select a relatively large number of commonly employed emoticons with and without clear emotional meanings to construct an emoticon space, where each emoticon serves as one dimension. Typically, the ESM consists of two phases: the projection phase and the classification phase. In the first phase, posts are projected into the emoticon space based on the semantic similarity between words and emoticons, which can be learned from a massive amount of unlabeled data. In the second phase, the assumption is made that posts with similar sentiments have similar coordinates in this space. Therefore, supervised sentiment classifiers are trained using the coordinates of the posts as features. By this means, sentiment analysis tasks, such as subjectivity classification, polarity classification and emotion classification can be well performed. Because the dimension of the emoticon space is considerably lower than the word space (i.e., the size of the vocabulary), the supervised classification tasks in emoticon space can be performed using less manually labeled data. Experimental results show that our simple model consistently outperforms previous state-of-the-art methods in various tasks.

2   Related Work

Both supervised and unsupervised methods have been used for microblog sentiment analysis. The construction of unsupervised classifiers based on existing sentiment lexicons is an option and labeled data are not required [3,19]. However, [3] determine that a classifier based on SentiWordNet [9] performs poorly compared with supervised methods. Other unsupervised methods based on emoticons or sentiment lexicons are proposed by [11,7]. [14,1] use sentiment lexicons as features to reinforce supervised classifiers.

Supervised methods are effective for microblog sentiment analysis [3,13]. [3] utilize the multinomial naive Bayes (MNB) and support vector machine (SVM) to perform supervised sentiment classification. First, several thousand training posts are manually labeled by a team of nine researchers. The limitation of supervised methods is that the performance is highly dependent on the size of the manually labeled data, which are always labor intensive to obtain.

Emoticons are frequently used to alleviate the problem. [20] use emoticons that have relatively clear emotional meanings as noisy labels of posts. They use ":)", and ":-)" as a noisy positive label and ":-(" and ":(" as a noisy negative label. Posts that contain these emoticons are used as training data, and a MNB classifier is trained for polarity classification. Similar work has been performed by [4] and [10]. [22] use emoticons as noisy labels to perform emotion classification on Sina Weibo. [11] verify sentiment indication of emoticons. Other signals, such as hashtags, have also been used as noisy labels [14,8]. We refer to these types of methods as noisy supervised methods for the remainder of this paper.

Additionally, noisy supervised methods are adopted to subjectivity classification. [20] regard the emoticons listed above (":)", ":(", etc.) as the noisy subjective label. To acquire noisy objective data, they assume that posts from accounts of popular newspapers are objective. In addition, [15] assume that posts containing an objective url link are objective.

Noisy supervised methods are negatively affected by the noises of training data. [15] propose a novel method to overcome the shortcomings of both supervised methods and noisy supervised methods.

3   Emoticon Space Model

In ESM, we use a number of frequently adopted emoticons to construct an emoticon space. As mentioned in previous section, ESM consists of two phases: the projection phase and the classification phase. For the projection phase, to satisfy our assumption that posts with similar sentiments have similar coordinates in the emoticon space, we obtain the coordinates of words using the semantic similarity between words and emoticons. Afterwards, the coordinates of the posts are obtained based on coordinates of words. For the classification phase, we use the coordinates of the posts as features for supervised sentiment classification tasks.

In this section, we first introduce a distributed representation of words [2], which provides an effective way to learn the semantic similarity between words and emoticons. Based on the distributed representation, we then present how words and posts are projected into an emoticon space. Finally, we introduce how the supervised sentiment classification is performed using the coordinates of the posts.

3.1   Distributed Representation of Words

The distributed representation of words has been widely used in neural probabilistic language models (NNLMs). In a distributed representation, each word is represented by a continuous real-value vector [2]. In this paper, words that are used in similar contexts are considered semantically similar and tend to have similar vectors [18]. The vectors can be learned by using a massive amount of unlabeled data and will be used later to project words into the emoticon space.

We use *word2vec* [17,16] to learn the distributed representation of words, because of its fast training speed.

3.2   Word Projection

While learning a distributed representation, we treat each emoticon as a unique special word. By leveraging large amounts of microblog corpora, the representation vectors of words are learned by *word2vec*. The representation vectors of all words (including emoticons) form a matrix $M_w \in \mathbb{R}^{d \times |V|}$, where $|V|$ is the size of the vocabulary and d is the dimension of the representation vectors. Each column of M_w denotes the representation vector of the corresponding word.

Suppose $|E|$ emoticons (denoted as $(e_1, e_2, ..., e_{|E|})$) are selected to construct the emoticon space. We search for the representation vectors of these emoticons in matrix M_w and receive a matrix $M_e \in \mathbb{R}^{d \times |E|}$. Each column in M_e denotes the representation vector of the corresponding emoticon.

In the distributed representation, words used in similar contexts tend to have similar vectors. Therefore, measuring the similarity between the representation vectors of word w_i and emoticon e_j helps identify their semantic similarity. In this paper, the cosine distance is used as the measurement of similarity between the representation vectors, which can be formalized as

$$similarity(\boldsymbol{w_i}, \boldsymbol{e_j}) = \frac{\boldsymbol{w_i} \cdot \boldsymbol{e_j}}{|\boldsymbol{w_i}| \, |\boldsymbol{e_j}|} \tag{1}$$

where $\boldsymbol{w_i}$ and $\boldsymbol{e_j}$ are the representation vectors of w_i and e_j. Specifically, if $w_i = e_j$, the similarity between the representation vectors is 1. We simply use equation 1 as the measurement of semantic similarity between w_i and e_j and use the semantic similarity as the coordinate of the word w_i in dimension j. Algorithm 1 shows the process of calculating the coordinate matrix of all words, which is denoted by C.

Algorithm 1. Calculation of word coordinates

Require:
    Distributed representation matrix of words, M_w
    Distributed representation matrix of emoticons, M_e
1: **for** each i in $[1 : |E|]$ **do**
2:    **for** each j in $[1 : |V|]$ **do**
3:       $C(i, j) = similarity(M_w(:, j), M_e(:, i))$
4:    **end for**
5: **end for**

In Algorithm 1, $C \in \mathbb{R}^{|E| \times |V|}$, and each column of C represents the coordinate of the corresponding word in the emoticon space. Thus, words have been projected into the emoticon space. If a word refers to an emoticon which is used to construct the emoticon space, then the corresponding column of C can be considered as the basis of this emoticon space. Because different emoticons are interrelated, this emoticon space is a non-orthogonal space.

3.3 Microblog Post Projection

In the last section, we have proposed a simple method to project words into the emoticon space using the semantic similarity between words and emoticons. The semantic similarity between posts and emoticons, however, can not be learned directly, but the coordinates of the posts can be obtained using basic mathematical operations on the coordinates of the words.. In this paper, we investigate two simple strategies for post projection.

Table 1. Statistics of the NLP&CC 2013 Benchmark Dataset

	neutral	like	happiness	sadness	disgust	anger	surprise	fear
training set	1823	597	371	388	425	235	112	49
testing set	4925	1525	1106	759	969	405	221	90
total	6748	2122	1477	1147	1394	640	333	139

– **Basic ESM (B-ESM)**

The simplest way to project a particular post into the emoticon space is to sum up the coordinates of the words that form the post. Formally, let p be the post, and $\boldsymbol{p}$ be the coordinate of the post. Therefore,

$$\boldsymbol{p} = \sum_{w_j \in p} C(:, j) \tag{2}$$

We name this strategy as Basic ESM, for its simpleness. Benefiting from the property that the coordinates of words are bounded in $[-1, 1]$, each word only has a limited effect for post projection.

– **Extended ESM (E-ESM)**

From our observation, many subjective posts contain one sentiment word or several sentiment words. Sentiment words may be semantically similar to some emoticons, and semantically dissimilar to other emoticons. For example, the word *happy* may be semantically similar to 😃, and semantically dissimilar to 😞. Thus, the coordinates of the word *happy* for the corresponding dimensions may be relatively large or small. Therefore, the maximum and minimum values of word coordinates in certain dimensions of a particular post may indicate the sentiment of this post.

ESM is flexible to integrate this information. Based on B-ESM, we add the maximum and minimum values above to the posts. Therefore, the coordinate of a particular post in each dimension can be denoted as a triplet of real values. The coordinate of the post can be formalized as:

$$\boldsymbol{p} = \begin{bmatrix} (\min_{w_j \in p} C(1, j), \sum_{w_j \in p} C(1, j), \max_{w_j \in p} C(1, j)) \\ (\min_{w_j \in p} C(2, j), \sum_{w_j \in p} C(2, j), \max_{w_j \in p} C(2, j)) \\ \dots \\ (\min_{w_j \in p} C(|E|, j), \sum_{w_j \in p} C(|E|, j), \max_{w_j \in p} C(|E|, j)) \end{bmatrix} \tag{3}$$

3.4 Supervised Sentiment Classification

After projecting posts into the emoticon space, the supervised sentiment classification tasks can be performed by using the coordinates of the posts as features. For B-ESM, the coordinates are used as feature vectors directly. For E-ESM, triples in all dimensions are concatenated to form a feature vector. The advantage of ESM is that emoticon signals are fully leveraged, and that it performs the supervised sentiment classification in an emoticon space in which the dimension is much lower than the word space. The next section will illustrate the performance of the two ESM models for the sentiment classification tasks.

4   Experiments

4.1   Experiment Setups

Experimental studies are performed on a public Chinese microblog benchmark corpus (NLP&CC 2013). This corpus consists of both a training set and a testing set The dataset is composed of fourteen thousand microblog posts collected from Sina Weibo and each of them is annotated with one of the following eight emotion tags: neutral, like, happiness, sadness, disgust, anger, surprise and fear. Posts in the dataset are not limited to certain topics. Table 1 shows details of the benchmark corpus. This dataset is adopted for the evaluation of polarity classification, subjectivity classification and emotion classification tasks. In polarity classification, emotions such as happiness and like are regarded as positive sentiment while sadness, anger, disgust and fear are regarded as negative sentiment In subjectivity classification, neutral is regarded as objective and the other seven emotion types are regarded as subjective. In emotion classification task, each emotion type serves as one class.

Table 2. Emoticons with clear emotional meanings

Sentiment	Example emoticons	Amount
positive		33
negative		19
happiness		9
like		10
sadness		7
disgust		5

To evaluate the performance of the proposed ESM model, 100 emoticons which are the most frequently used in microblog posts are selected to the construct the emoticon space. ICTCLAS [21] is adapted to the microblog corpus for Chinese word segmentation. LIBSVM [6] is chosen for supervised classification in ESM, and for baseline methods when SVM is needed. Dataset 1 is used for the projection phase of the ESMs.

Three state-of-the-art baseline strategies are adopted and compared with the proposed ESM framework for polarity, subjectivity and emotion classification tasks. Sufficient data with noisy labels are provided for the baseline methods and the parameters of the baseline methods are well tuned to achieve their best performance.

- **Supervised Methods:** These methods use manually labeled data to train classifiers. We investigate MNB and SVM, which are the most widely adopted classifiers for these tasks. We use binary feature vectors for SVM similar to [3]. We abbreviate supervised MNB and SVM as S-MNB and S-SVM, respectively.
- **Noisy Supervised Methods with Emoticons:** We use emoticons that have clear emotional meanings from the 100 emoticons mentioned above as noisy sentiment labels. Table 2 shows the five most frequently adopted emoticons for each sentiment. The assumption of [20] is adopted for the noisy objective label and outperforms the assumption of [15]. We implement a noisy MNB (i.e., N-MNB) classifier, which is adopted by most previous works.
- **Combination Methods:** [15] propose a state-of-the-art method that effectively integrates the supervised methods and the noisy supervised methods. Their method is called the ESLAM model. ESLAM is primarily used to identify polarity and subjectivity, but can be easily extended to perform emotion classification by taking emoticons that clearly express the corresponding emotions as noisy labels.

To better explore the benefits of differential treatment for each emoticon and the effect of emoticons that do not have clear emotional meanings, the E-ESM(*) and B-ESM(*), which use the same emoticons as the baseline methods, are investigated. The accuracy of all methods is reported to measure the performance of polarity, subjectivity and emotion classification tasks. Finally, we illustrate the improvement of the ESMs over the best runs in the NLP&CC 2013 benchmark.

4.2 Polarity Classification

For polarity classification, the original dataset has 3,599 positive posts and 3,320 negative posts. Next, a balanced dataset of 6,640 posts is randomly sampled, using the 3,320 negative posts and a random sample of 3,320 posts from the 3,599 positive posts. Similar to [15], a balanced subset of X total posts is randomly sampled for training along with 6,640 - X posts for testing. This procedure is performed for a given X and fixed parameters in ten rounds. The subjectivity classification and the emotion classification below use the same evaluation methodology. Classification accuracies for a different size of the manually labeled training set X are explored. In this task, X varies as 250, 500, 1000, 2000, 3000, 4000 and 5000.

In this task, the SVM with a linear kernel achieves the best performance for the ESMs. The polarity classification results of the ESMs and the baseline methods are compared in Figure 1. From this figure we can see that the B-ESM(*) and E-ESM(*) outperform all baseline methods, and indicate that the use of the same emoticons for ESM results in an improved utilization of emoticon signals. Emoticons that do not have clear meanings are useful when the number of training posts becomes relatively larger but have a negative impact when the training data are insufficient. The ESMs consistently outperform baseline methods for different training dataset sizes. According to the statistics, the E-ESM slightly outperforms the B-ESM because more features of the posts are introduced. When the training dataset size is larger than 4,000, the performance differences between the ESLAM and the supervised methods (S-MNB and S-SVM) become closer. However, a large gap still exists between the baseline methods and the ESMs.

Note that the ESMs are very robust even when the training size is small. The E-ESM with a training size of 250 outperforms all baseline methods with a training size of 5,000. For only a small number of manually labeled posts, a relatively high accuracy for the ESMs is obtained.

We then investigate the performance of the B-ESM and E-ESM using different numbers of emoticons. For a given number N, N emoticons that are most frequently adopted in Sina Weibo are selected for the ESMs to construct the emoticon space. Among the N emoticons, ones that have clear polarity indications are used as noisy sentiment labels for the ESLAM. The models are trained with 5,000 manually labeled posts. Figure 2 shows the performance of the models as N varies. We can see that with more emoticons adopted, the ESMs achieve a better performance and outperform the ESLAM. Moreover, the ESMs are flexible to leverage much more emoticons at no extra cost except for a few more computations.

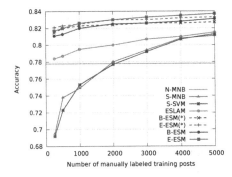

Fig. 1. Accuracies for different numbers of manually labeled training posts in polarity classification

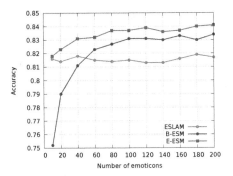

Fig. 2. Accuracies of the ESMs and the ESLAM using different numbers of emoticons in polarity classification

4.3   Subjectivity Classification

Similar to polarity classification, we randomly sample a balanced dataset of 13,496 posts for subjectivity classification. In this task, the size of the manually labeled training set varies as 250, 500, 1000, 2000, 4000, 6000, 8000 and 10000. The results are shown in Figure 3.

For baseline methods, when the training size is larger than 4,000, the S-SVM outperforms both the S-MNB and the ESLAM, and even slightly outperforms the B-ESM(*). However, the other three ESMs outperform all baseline methods

consistently, for different training sizes. All ESMs outperform the noisy supervised methods and the ESLAM. Meanwhile, after comparing the E-ESM with
the E-ESM(*) and the B-ESM with the B-ESM(*). We can see that emoticons
that do not have clear emotional meanings help subjectivity classification.

In subjectivity classification, the ESMs use more training data to obtain a
relatively high performance as compared with polarity classification. However, to
achieve a comparable accuracy, the B-ESM and E-ESM still require less labeled
data than the baseline methods.

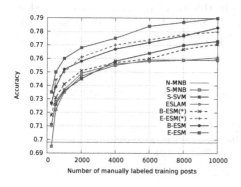
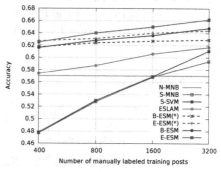

Fig. 3. Accuracies for different numbers
of manually labeled training posts in subjectivity classification

Fig. 4. Accuracies for different numbers
of manually labeled training posts in emotion classification

4.4   Emotion Classification

We use four emotion types (like, happiness, sadness and disgust) for a maximum
number of posts to perform this task because the quantities for the other three
types are relatively small. Similar to polarity classification, a balanced set of
4,588 posts is sampled and contains 1,147 posts for each type. The size of the
manually labeled training set varies as 400, 800, 1600 and 3200.

Figure 4 shows the results for this task. The results are similar to polarity
classification. the E-ESM outperforms the E-ESM(*) when the training size is no
less than 800 similar to the B-ESM. The ESMs outperform all baseline methods
for different training sizes and are robust and achieve high performances when
the training size is relatively small. For example, the E-ESM(*) achieves an
accuracy of 0.626 for 400 manually labeled training posts, which is higher than
the best performance (0.617) of the baseline methods with a training size of
3,200.

4.5   Comparison to Best Benchmark Results

The original dataset (NLP&CC 2013) was primarily used for an emotion recognition benchmark. To demonstrate the effectiveness of the emoticon space, we

use the ESMs to accomplish this benchmark and to compare its results to the best runs in this benchmark. By strictly following the benchmarking procedure, we use the training set in Table 1 for training and validation and the testing set for evaluation. For both the B-ESM and the E-ESM, we first use a subjectivity classifier to identify subjective posts. Next, these posts are classified by a seven-class emotion classifier.

The metrics for this benchmark include: 1) F1 score of the subjective class (the combination of seven emotion types, Subj F1); 2) Micro/Macro average of the F1 score for all emotion types (Micro F1, Macro F1) All results are illustrated in Table 3. The best runs for the metrics in this benchmark are abbreviated as BEST*. The superiority of the ESMs over BEST* is listed in parentheses. ESM achieves improvements of 8.1% on Subj F1, 24.7% on Micro F1 and 11.8% on Macro F1.

Table 3. Comparison between best runs of NLP&CC 2013 and ESMs

	Subj F1	Micro F1	Macro F1
BEST*	0.729	0.352	0.313
B-ESM	0.782(7.3%)	0.416(18.2%)	0.329(5.1%)
E-ESM	0.788(8.1%)	0.439(24.7%)	0.350(11.8%)

5 Conclusion

In this paper, we propose the emoticon space model (ESM) for microblog sentiment analysis. By differently treating each emoticon and integrating emoticons that do not have clear emotional meanings, ESM effectively leverages emoticon signals and consistently outperforms previous state-of-the-art methods.

Currently, post projection and supervised classification are two separated phases and labeled data do not improve the projection phase. In the future, we will investigate how to seamlessly integrate the two phases. Besides, different users may have different ways of using emoticons, which are worth studying.

References

1. Barbosa, L., Feng, J.: Robust sentiment detection on twitter from biased and noisy data. In: Proceedings of the 23rd International Conference on Computational Linguistics: Posters, pp. 36–44. Association for Computational Linguistics (2010)
2. Bengio, Y., Ducharme, R., Vincent, P., Jauvin, C.: A neural probabilistic language model. Journal of Machine Learning Research 3, 1137–1155 (2003)
3. Bermingham, A., Smeaton, A.F.: Classifying sentiment in microblogs: is brevity an advantage? In: Proceedings of the 19th ACM International Conference on Information and Knowledge Management, pp. 1833–1836. ACM (2010)
4. Bifet, A., Frank, E.: Sentiment knowledge discovery in twitter streaming data. In: Pfahringer, B., Holmes, G., Hoffmann, A. (eds.) DS 2010. LNCS (LNAI), vol. 6332, pp. 1–15. Springer, Heidelberg (2010)

5. Bollen, J., Mao, H., Zeng, X.: Twitter mood predicts the stock market. Journal of Computational Science 2(1), 1–8 (2011)
6. Chang, C.C., Lin, C.J.: Libsvm: A library for support vector machines. ACM Transactions on Intelligent Systems and Technology (TIST) 2(3), 27 (2011)
7. Cui, A., Zhang, M., Liu, Y., Ma, S.: Emotion tokens: Bridging the gap among multilingual twitter sentiment analysis. In: Salem, M.V.M., Shaalan, K., Oroumchian, F., Shakery, A., Khelalfa, H. (eds.) AIRS 2011. LNCS, vol. 7097, pp. 238–249. Springer, Heidelberg (2011)
8. Davidov, D., Tsur, O., Rappoport, A.: Enhanced sentiment learning using twitter hashtags and smileys. In: Proceedings of the 23rd International Conference on Computational Linguistics: Posters, pp. 241–249. Association for Computational Linguistics (2010)
9. Esuli, A., Sebastiani, F.: Sentiwordnet: A publicly available lexical resource for opinion mining. In: Proceedings of LREC, vol. 6, pp. 417–422 (2006)
10. Go, A., Bhayani, R., Huang, L.: Twitter sentiment classification using distant supervision. CS224N Project Report, Stanford, pp. 1–12 (2009)
11. Hu, X., Tang, J., Gao, H., Liu, H.: Unsupervised sentiment analysis with emotional signals. In: Proceedings of the 22nd International Conference on World Wide Web, pp. 607–618. International World Wide Web Conferences Steering Committee (2013)
12. Jansen, B.J., Zhang, M., Sobel, K., Chowdury, A.: Twitter power: Tweets as electronic word of mouth. Journal of the American Society for Information Science and Technology 60(11), 2169–2188 (2009)
13. Jiang, L., Yu, M., Zhou, M., Liu, X., Zhao, T.: Target-dependent twitter sentiment classification. In: ACL, pp. 151–160 (2011)
14. Kouloumpis, E., Wilson, T., Moore, J.: Twitter sentiment analysis: The good the bad and the omg! In: ICWSM (2011)
15. Liu, K.L., Li, W.J., Guo, M.: Emoticon smoothed language models for twitter sentiment analysis. In: AAAI (2012)
16. Mikolov, T., Chen, K., Corrado, G., Dean, J.: Efficient estimation of word representations in vector space. arXiv preprint arXiv:1301.3781 (2013)
17. Mikolov, T., Sutskever, I., Chen, K., Corrado, G., Dean, J.: Distributed representations of words and phrases and their compositionality. arXiv preprint arXiv:1310.4546 (2013)
18. Mnih, A., Hinton, G.E.: A scalable hierarchical distributed language model. In: Advances in Neural Information Processing Systems, pp. 1081–1088 (2008)
19. Nielsen, F.Å.: A new anew: Evaluation of a word list for sentiment analysis in microblogs. arXiv preprint arXiv:1103.2903 (2011)
20. Pak, A., Paroubek, P.: Twitter as a corpus for sentiment analysis and opinion mining. In: LREC (2010)
21. Zhang, H.P., Yu, H.K., Xiong, D.Y., Liu, Q.: Hhmm-based Chinese lexical analyzer ictclas. In: Proceedings of the Second SIGHAN Workshop on Chinese Language Processing, vol. 17, pp. 184–187. Association for Computational Linguistics (2003)
22. Zhao, J., Dong, L., Wu, J., Xu, K.: Moodlens: An emoticon-based sentiment analysis system for Chinese tweets. In: KDD, pp. 1528–1531. ACM (2012)

Predicting Health Care Risk with Big Data Drawn from Clinical Physiological Parameters

Honghao Wei[1,*], Yang Yang[1], Huan Chen[2],
Bin Xu[1], Jian Li[3], Miao Jiang[4,*], and Aiping Lu[4]

[1] Tsinghua University, Beijing, China
[2] Beijing University of Posts and Telecommunications, Beijing, China
[3] School of Basic Medical Sciences, Beijing University of Chinese Medicine, Beijing, China
[4] Institute of Basic Research of Clinical Medicine,
China Academy of Chinese Medical Sciences, Beijing 100700, China
O_HENRYWILL@126.com, miao_jm@vip.126.com

Abstract. Fatty liver often afflicts patients seriously and jeopardizes the health of human race with high possibility of deteriorating into cirrhosis and liver cancer, which motivates researchers to detect causes and potential influential factors. In this paper, we study the problem of detecting the potential influential factors in workplaces and their contributions to the morbidity. To this end, gender and age, retirement status and department information are chosen as three potential influential factors in workplaces. By analyzing those factors with demographics, Propensity Score Matching and classic classifier models, we mine the relationship between the workplace factors and morbidity. This finding indicates a new domain of discussing the causes of fatty liver which originally focuses on daily diets and lifestyles.

Keywords: Fatty liver, gender and age, retirement status, department information.

1 Introduction

Fatty liver disease (FLD), or simply fatty liver, is a very common chronic disease, whose prevalence continues to increase, especially in the "Western" world. According to the research by DeNoon (2013) [7], fatty liver disease occurs in 33% of European-Americans, 45% of Hispanic-Americans, and 24% of African-Americans. While large vacuoles of fat accumulate in liver cells via the process of steatosis, patients with fatty liver first suffer from an enlarged liver or vague right upper abdominal pains. Subsequently, in a subpopulation of patients, the disease progresses to more severe liver diseases, such as cirrhosis or total liver failure.[1] It eventually deteriorates into liver cancer, which damages tissues over time, significantly reduces the quality of life and ultimately leads to premature death.

Fortunately, fatty liver is reversible. If patients are diagnosed and treated promptly, the risk of the disease would decline significantly. In recent studies, alcoholic abuse,

---

* Corresponding authors.

H.-Y. Huang et al. (Eds.): SMP 2014, CCIS 489, pp. 88–98, 2014.
© Springer-Verlag Berlin Heidelberg 2014

together with obesity, are found to be the most influential factors. Though break-throughs have been made in the pathology of the disease, more undetected causes are there. They are hard to ignore since they can increase the incidence magnificently.

Currently, vast majority of passionate researches focus on the daily diets and life-styles. Previous researchers have found direct impacts of alcohol and obesity. Teli (1995) [15] reported the case of alcohol-induced fatty liver and suggested the hazard of excessive intake of alcohol. Ueno (1997) [17] noticed the increasing incidence of obese patients with fatty liver and proposed restricted diet and exercise therapy as useful methods to combat the disease. Similar work has been done by Tominaga (1995) [16], who elicited a direct relationship between the degree of obesity and fatty liver in children.

Influence analysis has attracted considerable research interests and is becoming a popular research topic (Wanless 1990[18], Angulo 2002[3], Marchesini 2003[12], Fabbrini 2010[8],). However, most exciting works focused on daily diets and life-styles. On the contrary, the influential factors in workplaces have been largely ig-nored. Since the working hours occupy approximately half of an adult's time, it is reasonable to involve workplace factors in the exploration of causes of fatty liver.

In this paper, we aim to quantitatively study the influential factors in workplaces and how they affect the incidence of fatty liver. Our objective is to effectively and efficiently discover the underlying influence pattern related to workplace.

The rest of the paper is organized as follows: Sect.2 formally formulates the problem; Sect.3 proposes relevant methods; Sect.4 introduces evaluation aspects and classic classifier models, and presents experimental results validating our assump-tions; Sect.5 displays new discoveries and offers the potential explanations; Sect.6 concludes.

2   Problem Definition

In this section, we define several relevant concepts in order to formulate the fatty liver prediction problem.

The following concepts are introduced:

2.1   Definition 1. Lab Test Records

We define the lab test record of patient n to be $X_n = \{X_I\}$ where X_I denotes a lab test performed on the patient n. We also use set $\{a_1, a_2, \dots, a_{46}\}$ to represent the lab test items evaluating the health condition of the patient, and for each patient we final-ly choose 46 major items to denote their health status among a range of alternative attributes. Therefore, we have $X_I = \{a_1, a_2, \dots, a_{46}\}$. We denote Y_n as a label node that indicates whether the patient n has fatty liver.

In our experiment, we use the classical models roughly to evaluate the potentially relevant lab test items and select 46 items of them to be the attributes to forecast the results. The selected attributes are listed as follows in table 1:

We screen fatty liver cases from people in several stable organizations of Chaoyang District, Beijing, China. The data are obtained from their physical examinations. Meanwhile, a patient could have several lab tests in different organizations and the results could be different. While we could deduce the differences of health conditions of the patient, the lab tests from different organizations with different features and instructions could make a difference. In order to guarantee the integrity of the data and avoid extra turbulence of different tests, we acquire the data of four years from one single medical organization. In addition, all the people are instructed to receive all the lab test items so that our 46 attributes are ensured to exist in the test report.

Table 1. Attribute in test

	Gender	Age	Systolic Pressure	Weight	Height
	BMI	Pro	Diastolic Pressure	Thyroid	LYM
	MONO	NEUT	LYM#	MONO	NEUT
	RBC	MCV	Hematocrit	MCH	MCHC
	PLT	MPV	PCT	PDW	RDW
Attribute	HGB	DB	GLU(UR)	KET	SG
	OB	pH	Waistline	URO	NIT
	WBC	ALT	Cr	URIC	GLU
	CHO	TG	Gallbladder	AST	HDL
	Kidney				

Consequentially, we select the lab test results from 2010 to 2013 and utilize our proposed models to forecast the prevalence and uncover its potential relationship with influential factors in workplaces.

2.2 Definition 2. Fatty Liver

In this paper, we define fatty liver as the condition in which large vacuoles of triglyceride fat accumulate in liver cells. This is usually accompanied by defect of the ability to respond to metabolic stress. With the impairment of this ability, a number of diseases and disorders can be incurred.

All the patients diagnosed with fatty liver have undertaken physical exams, blood tests and ultrasound examinations. Liver enzymes higher than normal level can be found and fat in the liver has been detected in a computed tomography scan or magnetic resonance imaging test. In clinical practice, patients with these clinical traits are also considered as having the disease and we adopt the same criterion in our study[11].

3   Approach

In this section, we briefly introduce the Logistic Regression and Propensity Score Matching.

3.1   Logistic Regression

Logistic Regression is a type of probabilistic statistical classification model. Considering p independent variables $x = \{x_1, x_2, ..., x_p\}$, we define conditional probability $P(Y = 1|x) = p$ as the possibility of occurrence based on measurement. The explanation of Logistic Regression begins with the possibility of occurrence. We define it as follows:

$$P(Y = 1|x) = \frac{1}{1 + e^{-g(x)}}$$

Where $g(x) = \beta_0 + \beta_1 X_1 + \beta_2 X_2 + ... + \beta_p X_p$

The possibility of non-occurrence refers to be:

$$P(Y = 0|x) = 1 - P(Y = 1|x) = 1 - \frac{e^{g(x)}}{1 + e^{g(x)}} = \frac{1}{1 + e^{g(x)}}$$

In this way, the odds of experiencing an event (the odds of the possibility of occurrence and non-occurrence) is

$$\frac{P(Y = 1|x)}{P(Y = 0|x)} = \frac{p}{1 - p} = e^{g(x)}$$

The logarithmic term of the above is the linear function that we demand.

In case we have n observation samples, the observed values refer to $\{y_1, y_2, ..., y_n\}$. The probability of occurrence could be written as

$$y_i = p_i^{y_i}(1 - p_i)^{1 - y_i}$$

Since all observations are independent, we get the likelihood function

$$l(\beta) = \Pi_{i=1}^n p_i^{y_i}(1 - p_i)^{1 - y_i}$$

Using Maximum likelihood estimation, such as Newton-Raphson's method [21], we determine the coefficient before each variable.

3.2   Propensity Score Matching

In order to estimate the influence of a single attribute, we create a dynamic matched sample of treated and untreated nodes. The treated and untreated groups feature differently in the targeted attribute. Bias might arise because of the apparently different outcomes of these two groups of units. In experiments, the randomization enables unbiased estimation of treatment effects by the law of large numbers. However, unfortunately, in observational studies, the assignment of treatments to research subjects is not randomized. For the purpose of eliminating the bias between treated and untreated units and narrowing down the differences, we adopt Propensity Score Matching. Full details regarding Propensity Score Matching methods are provided in Caliendo's (2008) [5] guidance for the implementation.

In the implementation of Propensity Score Matching, we match treated nodes with untreated ones that are likely to have the same attributes, such as similar health condition etc. For every examinee (we label him or her as the t th examinee in the treated group), we estimate pit, the propensity to have been treated, using a logistic regression of the likelihood of health status, which is conditional on a vector of observable characteristics and clinical traits (X). It is listed as follows:

$$p_{it} = P(x_{it} = 1 \mid X_{it}) = \frac{e^{g(x)}}{1 + e^{g(x)}}$$

Where $g(x) = \beta_0 + \beta_1 X_1 + \beta_2 X_2 + \ldots + \beta_p X_p$

We drop matched pairs of which the distance of propensity scores exceeds two standard deviations. For all treated examinee i, we choose an untreated match j such that $\left\| p_{it} - p_{jt} \right\|$ is minimized to conform to $\min \left\| p_{it} - p_{jt} \right\| \leq 2\,\sigma_d$ where σ_d is the standard deviation of the distance $\left\| p_{it} - p_{jt} \right\|$. We then compare fractions of treated (n+) and untreated (n-) adopters. Once the odd exceeds integer 1, the targeted attribute is believed to have impacts on the final result [4], [14].

4   Experiment

We use a collection of real medical records from a famous hospital in Beijing. The data set spans four years, containing 5535 medical records corresponding to 1985 individual patients and 91 kinds of lab tests in total.

We view each examinee's record as an instance, and aim to infer whether the corresponding indicators have potential effects on and contributions to fatty liver. Three influential factors are taken into account in our experiments: gender and age, retirement status, and department information.

We use demographic methods to demonstrate the influence of age and gender. In the meantime, for the purpose of good matching, Propensity Score Matching is proposed to evaluate the influence of retirement status. Last but not least, we adopts several classic classifier models to test the department attribute.

4.1   Evaluation Aspects

We evaluate our methods from the following two aspects:

Precision and Recall. In pattern recognition and information retrieval with binary classification, precision is the fraction of retrieved instances that are relevant, while recall is the fraction of relevant instances that are retrieved.

F-measure. The F-measure can be interpreted as a weighted average of the precision and recall, where an F-measure reaches its best value at 1 and worst value at 0. The f-measure is defined as:

$$F_1 = 2 \times \frac{Precision \times Recall}{Precision + Recall},$$

We compare the results of fatty liver forecast generated by following methods:

Logistic Regression. Lab test results are treated as features and LR (Alison 1999) [2] is employed as the classification model for fatty liver disease forecasting.

Naive Bayes. A traditional Naive Bayes model (Frank, Trigg, Holmes, Witten 2000) [10] is used for prediction.

J48 Decision Tree. C4.5 Decision Tree is considered as well. For convenience, we employ the J48 Decision Tree provided by Weka 3.6 (Frank 2004) [9].

All algorithms are implemented in Java and C++, and all experiments are performed on a Lenovo laptop running Win 7 with AMD 1.4 GHz and 4 GB of memory.

4.2   Gender and Age

We imply the potential differences among distinct age and gender groups on the possibility of getting fatty liver. In our demographic test, we select 5535 instances to demonstrate the relationship. Among them, male examinees and female examinees are approximately half-and-half, which guarantees the representativeness and universality of the data.

By way of eliminating extreme samples, we set a series of criterion to screen the data (see Table 2). All selected female and male participants should meet the following standards in Table 2.

The demographic results are showed in Fig.1 and Fig.2. With these figures, we infer that age factor shares linear relationship with the incidence of fatty liver. Therefore, we analyze the data with linear fitting. For female participants, the correlation coefficient is 0.9221. In contrast, the correlation coefficient of the male participants is only 0.7081. The large gap between the correlation coefficients of two gender groups needs further discussion.

Table 2. Criterion for examinee enrollment

Clinical item	Female	Male
Systolic Pressure	>= 110 and <= 140	>= 110 and <= 140
Diastolic Pressure	>= 70 and <= 90	>= 70 and <= 90
Weight	>= 45 and <= 75	>= 60 and <= 80
Height	>= 155 and <= 175	>= 165 and <= 185
BMI	>= 15 and <= 25	>= 19 and <= 25
Blood Sugar	<= 6	<= 6
Cholesterol	<= 6	<= 6
Glycerol acid	<= 2	<= 2

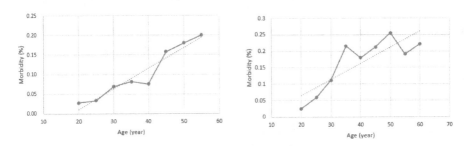

Fig. 1. Morbidity and Age (female participants (left) and male participants (right))

4.3   Retirement Status

In the interests of large number laws, the number of unretired and retired employees involved in the matching exceeds 800. Additionally, we utilize Propensity Score Matching to create reasonable pairs of the retired and unretired with similar health conditions and limited age differences.

Fig. 3 demonstrates the result of Propensity Score Matching. Obviously, all the final odds each year exceed integer 1. Though the odds fluctuates due to the different data each year, it is still convincing that retirement status contributes to prevalence of fatty liver significantly.

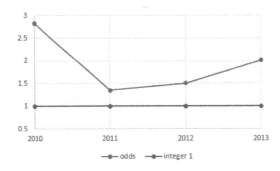

Fig. 2. The odds of Propensity Score Matching in different years

4.4   Department Information

The demographics of the rate of fatty liver suggest the potential relationship between career and the prevalence of illness. To illustrate the situation, 7 typical departments have been chosen and the corresponding morbidity has been calculated.

From table 3, the differentiation of morbidity among departments has been evidently displayed. The incidence of Human Resources Department is five times as much as that of Research and Development center.

Table 3. Incidence of fatty liver in different departments

Department	Incidence
Technique Department	31.25%
HR Department	61.36%
Machining Department	31.19%
Administrative Department	27.86%
Finance Department	26.32%
Express Mail Services	23.08%
Research and Development Center	13.79%

To further clarify the influence of careers and working departments, we randomly take HR Department and Technique Department as an instance. Three classic classifier models are used to test the examples. They are Logistic Regression, Naive Bayes and J48 Decision Tree.

Table 4 confirms the theory and all these three models exhibit that applying career and department information as attributes would increase the F1-measure. The most apparent improvement of the performance rests in Logistic Regression. The F1-measure increases up to 14.16%.

Table 4. Prediction results about department information

Class	Method	Precision	Recall	F1
With	LR	0.650	0.648	0.629
Department	Naive Bayes	0.636	0.639	0.636
Information	J48	0.584	0.593	0.561
Without	LR	0.561	0.571	0.551
Department	Naive Bayes	0.616	0.620	0.615
Information	J48	0.560	0.574	0.547

5   Discussion

In this section, we mainly deal with the results in the experiment part. Through further analysis, we draw the conclusion on the relationship between the workplace factors and morbidity.

5.1   Gender and Age

The linear relationship displayed in Fig. 1 and Fig. 2 clarifies the potential contributions of gender and age to the morbidity of fatty liver. We notice that the morbidity increases with age. It conforms to the regular senescence patterns of human beings.

However, there are three more interesting questions worth noting: 1) Why the correlation coefficient of male examinees are much lower than that of the female examinees; 2) Why the apparent skewing of female and male participants occurs at different ages; 3) Why at those ages, the incidence of fatty liver within the female units inclines to decrease while the incidence within male units increases sharply?

To better answer these questions, we need to understand the different career choice and social roles of different genders. In the sample selected, male participants tend to take career more physically and psychologically demanding, such as manual work or scientific research. Female participants, on contrary, usually work in civil services, such as accountants. The pressure of the career differs and the external interferences are more likely to disrupt the male workers' original internal senescence patterns, thus resulting in the decline of correlation coefficient. The obesity status recognizes this assumption. Table 5 illustrates the BMI categories for male and female. It is worth noting that much higher percentage of male participants (46.41%) than female (21.79%) are overweight or even obesity. It is ultimately consistent with fatty liver studies since obesity is one of the major causes of the illness (Centis 2013) [6].

Table 5. BMI categories for male and female

Class	Male	Female
Underweight (BMI ≤18.5)	2.65%	7.69%
Normal weight (BMI = 18.5 ~ 24.9)	50.95%	70.51%
Overweight (BMI = 25 ~ 29.9)	40.23%	18.91%
Obesity (BMI ≥30)	6.18%	2.88%

To be more convincing, we analyze the important timing point and age bracket of these two units. For women, age 40 is a watershed. The incidence of fatty liver in the early 40s is significantly under prediction by linear regression while during late 40s the prevalence soars high. We deduce that, by the age of 40, women have fulfilled the responsibility about marriage and conception. The era of promoting in career has passed as well (O'Brien 1993) [13]. Therefore, the 40s women tends to be more relaxed, tolerant and resilient which aid to confront the metabolic diseases such as fatty liver. Additionally, the physical condition of women deteriorates swifter than men. It usually takes place at the age of 40s when obesity begins to increase. Clinical study has also proved that the fatty liver peak for women is usually 40s with endocrine disorders (Yan 2013) [20]. Therefore, the rate of fatty liver is raised as expected in the late 40s.

For male participants, the sudden rush of morbidity takes place at age 35 and 50. They are the important preparation periods for man: one for the golden development era in 40s, another for the last chance of promotion in 50s. The relevant performances would be irregular and greasy diet, insufficient rest time and work in overload. It is found in clinical practices that the intake of edible oil and alcohol increases while the sleeping hour decreases significantly during the period (welsh 2012) [19]. This suits the career pattern of a man and strengthens the relationship between the morbidity and workplace factors.

5.2   Retirement Status

By comparing the senior citizens with similar age and health condition, yet with different retirement status in Propensity Score Matching, the contribution of retirement status becomes obvious. The odds of data in four years are larger than the integrity 1.

It is noticed that those examinees who retire at their 60's usually occupy the important position in workplaces in the last years of their careers. They are either the leader or the person in charge. The pressure and demands, which probably account for the temporary decline of morbidity, varies greatly before and after retirement.

5.3   Department Information

To paraphrase the effect of department information, we need to take these elements into consideration: gender rate, average age and the specialty of career. Firstly, the specialty of career distinguishes the morbidity in different professions. For instance, the department with highest incidence of fatty liver are HR department and machining department. Both of these jobs are physically demanding and exhausting. What's more, the sex rate and average age directly make accounts in tackling the problem. The average age in Research and Development Center is 32 while the average age in Machining is 39. The percentage of female employees in Machining is below 14% while the percentage in Administrative Department is up to 90%. From the last two parts, we conclude that women have comparatively lower morbidity than men and the rate increases with age. Therefore, the situation in different departments differs.

6   Conclusion

In the framework of data analysis, this paper aims to investigate the potential causes of fatty liver in workplaces regarding gender and age, retirement status and department information. Data from 5535 instances suggests the potential relationship between the influential factors in workshop and the prevalence of fatty liver. The skewing from linear relation of female in early 40s and of male in 35 and 50s suits the career pattern of two units. Additionally, the subtle effect of retirement status on decreasing the morbidity temporarily has been detected. Furthermore, we explore the different morbidity in distinct departments and the situation is in line with our assumption.

Based on these findings, in order to improve the prediction of the disease and take precautions, female employees in late 40s and male in 35 and 50s should be suggested to take regular physical examination and regular life pattern is needed. Moreover, seniors before the retirement age should be instructed to avoid work overload and learn to release stress. Last but not least, the health condition of particular departments should be paid special attention and the employers should be provided with more opportunities for body check and rest.

Regarding some limitations of this study, further research is necessary.

First, this study aims to find the causes in workplaces. To confirm the conclusion of this study, further studies are needed to investigate more factors especially those unique ones in workplaces.

Next, the relationship among the influential factors has not yet been discussed. It is unclear what the most influential factors is in workplaces and whether some of the factors would affect others (e.g. the influence of gender differences on retirement status regarding the morbidity).

At last, the targeted solutions and their impacts should be explored and examined. Further guidelines for early warning and diagnosis are needed as well, to develop a better understanding of fatty liver and share this with others.

References

1. Adler, M., Schaffner, F.: Fatty liver hepatitis and cirrhosis in obese patients. The American Journal of Medicine 67(5), 811–816 (1979)
2. Allison, P.D.: Logistic regression using the SAS system. SAS Institute. Inc., Cary (1999)
3. Angulo, P.: Nonalcoholic fatty liver disease. New England Journal of Medicine 346(16), 1221–1231 (2002)
4. Aral, S., Muchnik, L., Sundararajan, A.: Distinguishing influence-based contagion from homophily-driven diffusion in dynamic networks. Proceedings of the National Academy of Sciences 106(51), 21544–21549 (2009)
5. Caliendo, M., Kopeinig, S.: Some practical guidance for the implementation of propensity score matching. Journal of Economic Surveys 22(1), 31–72 (2008)
6. Centis, E., Moscatiello, S., Bugianesi, E., et al.: Stage of change and motivation to healthier lifestyle in non-alcoholic fatty liver disease. Journal of Hepatology 58(4), 771–777 (2013)
7. De Noon, D.J.: Fatty liver disease: Genes affect risk (2013)
8. Fabbrini, E., Sullivan, S., Klein, S.: Obesity and nonalcoholic fatty liver disease: biochemical, metabolic, and clinical implications. Hepatology 51(2), 679–689 (2010)
9. Frank, E., Hall, M., Trigg, L., et al.: Data mining in bioinformatics using Weka. Bioinformatics 20(15), 2479–2481 (2004)
10. Frank, E., Trigg, L., Holmes, G., et al.: Technical note: Naive Bayes for regression. Machine Learning 41(1), 5–25 (2000)
11. Liver, F.: Guidelines for diagnosis and treatment of nonalcoholic fatty liver diseases. Chinese Journal of Hepatology 3 (2006)
12. Marchesini, G., Bugianesi, E., Forlani, G., et al.: Nonalcoholic fatty liver, steatohepatitis, and the metabolic syndrome. Hepatology 37(4), 917–923 (2003)
13. O'Brien, K.M., Fassinger, R.E.: A causal model of the career orientation and career choice of adolescent women. Journal of Counseling Psychology 40(4), 456 (1993)
14. Peikes, D.N., Moreno, L., Orzol, S.M.: Propensity score matching. The American Statistician 62(3) (2008)
15. Teli, M.R., James, O.F.W., Burt, A.D., et al.: The natural history of nonalcoholic fatty liver: A follow-up study. Hepatology 22(6), 1714–1719 (1995)
16. Tominaga, K., Kurata, P.J.H., Chen, Y.K., et al.: Prevalence of fatty liver in Japanese children and relationship to obesity. Digestive Diseases and Sciences 40(9), 2002–2009 (1995)
17. Ueno, T., Sugawara, H., Sujaku, K., et al.: Therapeutic effects of restricted diet and exercise in obese patients with fatty liver. Journal of Hepatology 27(1), 103–107 (1997)
18. Wanless, I.R., Lentz, J.S.: Fatty liver hepatitis (steatohepatitis) and obesity: An autopsy study with analysis of risk factors. Hepatology 12(5), 1106–1110 (1990)
19. Welsh, J.A., Karpen, S., Vos, M.B.: Increasing prevalence of nonalcoholic fatty liver disease among United States adolescents, 1988-1994 to 2007-2010. The Journal of Pediatrics 162(3), 496–500, e1 (2013)
20. Yan, J., Xie, W., Ou, W., et al.: Epidemiological survey and risk factor analysis of fatty liver disease of adult residents, Beijing, China. Journal of Gastroenterology and Hepatology 28(10), 1654–1659 (2013)
21. Jennrich, R.I., Sampson, P.F.: Newton-Raphson and related algorithms for maximum likelihood variance component estimation. Technometrics 18(1), 11–17 (1976)

Gender Identification on Social Media

Xiaofei Sun, Xiao Ding, and Ting Liu

Research Center for Social Computing and Information Retrieval,
Harbin Institute of Technology, Harbin, China
{xfsun,xding,tliu}@ir.hit.edu.cn

Abstract. Accurate identification of hidden demographic attributes from social media is very useful for advertisement, personalized recommendation and etc. We investigate the effect of two different classification models for the gender identification problem over different attributes of Sina Weibo users. To improve the accuracy of the classfication models, we propose a novel feature selection algorithm and a retrained multiattribute model. Experimental results show that the accuracy of our approach achieves 89.01% which is better than any previous work in this problem.

Keywords: Gender identification, Social media, Retrained Multi-attribute.

1 Introduction

In recent years, online social networks like Facebook, Twitter and Sina Weibo have grown impressively, which allow millions of users share their daily life through texts, photos and other multimedia. However, a common problem is that its easy to provide a false information or just hide it, which, definitely, makes it hard for advertising and may even lead to cyber crimes. Therefore, accurate identification of these attributes has attracted a great deal of research interest, especially in gender identification and age identification [5,8].

In this paper, we want to identify the gender attribute on Sina Weibo. We treat the gender identification as a binary classification problem and investigate two classification models with three feature selection algorithms for it.

The rest of this paper is organized as following. Section 2 briefly describes some releted work and compares the work of Burger et al. (2011)[1] and Peersman et al. (2011)[6] with ours, pointing out the differences. In Section 3, we describe the feature selection algorithms and retrained multi-attribute model used in our experiments. And Section 4 describes the data set. The experiments and results are introduced in Section 5. We pay attention to the changing of accuracy with the size of training data by using different classifiers, feature counts and feature selection algorithms. We also research the effect of the size of training data and the retrained multiattribute model for final accuracy. Finally, Section 6 summarizes the findings and introduces the future plans.

2 Related Work

There are several studies regarding gender identification on social media[3,4,7]). Burger et al. (2011)[1] use 146,925 words on average as training data to discriminate gender

H.-Y. Huang et al. (Eds.): SMP 2014, CCIS 489, pp. 99–107, 2014.

on twitter. The best accuracy of their approach is 92%. One of the most useful attributes on their experiments is user's $FullName$, which must be true name. However, it can be much more difficult to apply the same method to Chinese users, because it's much harder to distinguish Chinese gender purely through their names (compared to english names) and users of Sina Weibo needn't fill out their true name.

Moreover, Peersman et al. (2011)[6] study the relationship between gender and age. Their experiments, relating gender with age, produce a best accuracy of 92%. However, most users of Sina Weibo will not fill out the information of age or just provide a false age. In addition, the best accuracy can be obtained only from a certain age bracket. Our experiments, on the contrary, do not rely on the age attribute at all.

3   Approach

3.1   Feature Selection

If we simply use bag-of-words feature, we have to face the curse of dimensionality (more than 100,000,000) and the problem of sparseness feature. So we use three different feature selection algorithms to solve this problem and intestigate the effect of different feature selection algorithms for gender identification problem. We use TF-IDF [9] and CHI [9] algorithms in experiment and propose another better algorithm for this problem.

- **TF-IDF:** It's a numerical statistic that is intended to reflect how important a word is to a document in a collection or corpus.
- **CHI:** CHI is a popular feature selection method using Chi-squared test in text classification problem. However, there is a big disadvantage that CHI will give more weight to words with low frequency. For example, if the sentence is AAAAAAAABCCC, the weight of A, B and C get the same value of CHI. However, word A should carry more information of gender than B and C because user use A more ofen.
- **Chi-square test with cosine similarity (COS-CHI):** To solve the problem of CHI, we use cosine similarity to evaluate the weight of a word in CHI. For a certain attribute, we define some symbols as Table 1 to caluculate the COS-CHI of word w. And using these symbols, we define weight of word ($weight_w^i$) and COS-CHI (χ_{cos}^2) as Equation 1 and Equation 2.

$$weight_w^i = \frac{n_w^{i\,2}}{length_i} \tag{1}$$

$$\chi_{cos}^2(w) = \frac{N(AD - BC)^2}{(A+C)(A+B)(B+C)(B+D)} \tag{2}$$

Table 1. Mathematical Symbols

Symbol	Explain
N	the number of total users
g_i	the gender of $user_i$; 0 means male, and 1 means female
n_w^i	the number of word w in user i
l_i	the attribute size user i
$weight_w^i$	the weight of word w in user i
A	the sum of $weight$ for word w in users whose gender is g
B	the sum of $weight$ for word w in users whose gender is $1 - g$
C	the count of users whose gender are g
D	the count of users whose gender are $1 - g$

Table 2. User Attributes

Attribute	Meaning	Example
$ScreenName$	An unique name created by users when they register	Guo Yunyan
$Emoticon$	The emoticons used in tweets	[very happy],[embarrassed]
$Birthday$	User's birthday	1991/12/03
Job	User's job	College Student
$Constellation$	User's constellation	Sagittarius
Tag	Several tags set by user to show user's interest or characteristics	[Computer], [Fan of Constellation], [Life]
$Tweets$	Tweets published by user	Cappuccino always brings so much happiness!
$Description$	Description set by user to show user's state and it's more stable than tweets	Why has no one told me that my hand would be cut if I still use microblog?

3.2 Retrained Multi-attribute Model

We combined the results of different attributes classfier together, and then retrain it.

We paid special attention to three attributes:$UserPost$, $Emoticon$ and $Screen$ $Name$. For these three attibutes, we chose the best classfiers for them and get three probabilities which conflects the probabilities of that user is male. Then this problem changes into a classfacation problem in a three dimensions. So we use SVM classfier to retrain it and use this retrained model as the final classfier.

4 Data

We chose Sina Weibo as our data source which is the biggest online social network in China. And we chose verified users as our training data whose infomation has been verified by Sina Weibo. We crawl the information of about 200,000 users randomly from hundreds of different professions. To ensure the reliability of data, we only use the

Table 3. Common Illegal Postfix

Illeagal Postfix
Official Website
Public Microblogging
Travel Agent
Publisher
Corporation
Platform
Forumd
University
Take-away
Association

data of active users whose total amount of posts is larger than 50 and have registered more than half a year. Moreover, active user has to submit at least one post in last week. For each active user, we crawl at least 50 user posts.

And there are lots of structured user profiles for Sina Weibo users and they are shown in Table 2.

5   Experiments

We treat the gender identification problem as a binary classification problem, i.e. male or female. We pay attention to five user attributes: $ScreenName$, $Description$, Tag, $Emoticon$ and $UserPost$.

The real percentage of male users in training data is 57%, and we use this value as the benchmark.

5.1   Pre-processing of the Data

The data pre-processing contains three steps.

- **Step 1:** Remove the public accounts or machine users. The data of 200,000 users contains lots of public account such as offical website or other public home pages. To clean the data, we use a simple strategy with a list of illegal postfix. Some common postfix are shown in Table 3. User will be judged as illegal user if the screen name contains these postfix .
- **Step 2:** Remove inactive users which are mentioned in Section 3.
- **Step 3:** Parse the word. We used LTP-Cloud[2] to parse the text.

Finally, we get the data of about 36,000 users as the experimental data.

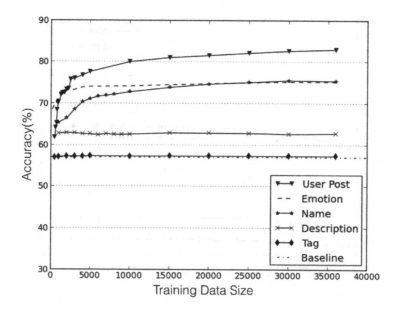

Fig. 1. SVM: Accuracy with Train Data

Table 4. Best Accuracy of Each Attribute

Attribute	Naive Bayes	SVM
$ScreenName$	77.53%	75.44%
$Description$	63.03%	62.77%
Tag	59.44%	57.36%
$Emoticon$	77.97%	75.26%
$UserPost$	75.72%	83.01%

5.2   Experiments on Single User Attribute Using Two Algorithms

Using two classification models, Naive Bayes and SVM, we test the accuracy of each attribute with the increasing of training data size. Moreover, for each model, we test the effect of different feature selection algorithms, which contain TF-IDF, CHI and COS-CHI.

The results of single attribute with SVM and Naive Bayes is shown in Figure 1 and Figure 2. And the highest accuracy of each attribute are shown in Table 4.

In SVM, it's obvious that the performance of $UserPost$ is the best and the accuracy is at least 5% higher than other attributes.

The performance of $ScreenName$ and $Emoticon$ are as good as each other in big data size ($size \geq 20000$) and the highest accuracy are 75%. But in small one ($size <$ 20000), the result of $Emoticon$ is better than $ScreenName$. When the data size is smaller than 1000, $Emoticon$ even performs better than $UserPost$. That is caused by differant feature size of these attributes. The feature size of $UserPost$ is 8000 while

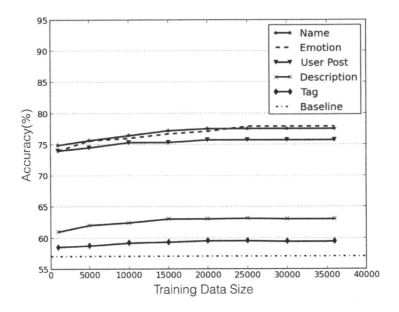

Fig. 2. Naive Bayes: Accuracy with Train Data

that of *Emoticon* is only 1400, so *UserPost* needs more training data to get a good result.

The highest accuracy of attribute *Description* is 62%. And that of *Tag* is only 57.36%, which is just little higher than the benchmark (57%). Because only 23.67% people will set tags about themselves and it's usually no more than 4 words so it's difficult to get gender information.

As for Naive Bayes, the best performance comes from attribute *Emoticon* with an accuracy as high as 77.97%. Performance of attribute *ScreenName* is as good as *Emoticon* getting an accuracy of 77.53%. At the same time, except attribute *UserPost*, accuracy of other attributes are all a little higher than that of SVM.

For all five attributes, the accuracy increase with training data size using this two models. Accuracy of Bayes in all five attributes is limited below 77%. On the contrary, accuracy of SVM reaches 83% when data size exceeds 36,000 in attribute *UserPost*. But Naive Bayes is more accurate and steady than SVM in small data size.

5.3   Experiments with Different Sizes of User Attribute

We also want to investigate the effect of the size of test data for the final accuracy, because the length of these data varies from user to user and some attributes may be empty.

We test this effect only in attribute *Emoticon* and *UserPost*, for the size of remaining attributes is very small and cannot get valuable conclusion.

As shown in Figure 3, the accuracy increases with the increasing of the test data. And the accuracy of *Emoticon* is higher than *UserPost* when *size* < 90, and as large as

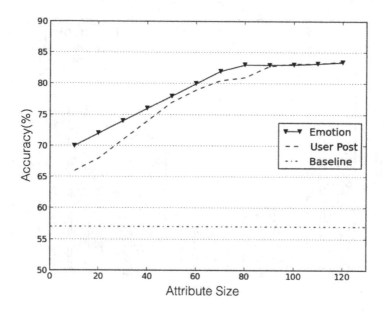

Fig. 3. Accuracy with Feature Count in SVM

$UserPost$ when $size \geq 90$. The result prove that the using pattern of emotions carries more information about user's gender. However, some users may use few emotions and that's why the accuracy of $Emoticon$ is 8% smaller than that of $UserPost$.

5.4 Experiment with Different Feature Selection Algorithms

The result shown in Figure 4 indicates that TF-IDF is the worst feature selection algorithm for all attributes.

And for attribute $ScreenName$, Tag and $Description$, the performance of CHI is as good as COS-CHI. As we mentioned in Section 5, the aim of COS-CHI algorithm is to make up for the disadvantage of CHI which will give more weight to words with low frequency in a certain attribute. In these three attributes, it's unlikely to appear repetitive words, which means the frequency of a word would only be 1 or 0. So there is no difference between the performances of CHI and COS-CHI. However, for other attributes such as $Emoticon$ and $UserPost$, COS-CHI is better than CHI. It's because that there are lots of repetitive words in this three attributes, especially in $UserPost$ and $Emoticon$.

5.5 Experiment with Retrained Multi-attribute Model

The result of retrained multiattribute model shown in Figure 5 indicates that the accuracy of retrained multiattribute model is better than any sigle attribute model, even when training data size is small. And the best accuracy 89.01% is also got from this model.

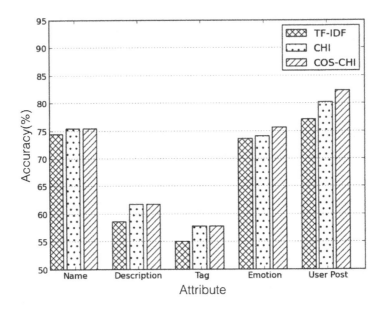

Fig. 4. Accuracy with Different Feature Select Algorithm

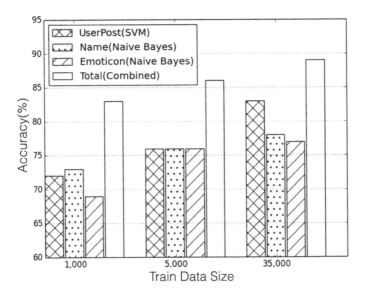

Fig. 5. Accuracy with Retrained Model

6 Conclusion

In this paper, we have presented several configurations of two classifiers for identifying the gender of Sina Weibo users and propose a novel feature selection algorithm using cosine similarity and a retrained multi-attribute model. The large dataset used for construction and evaluation of these classifiers is crawled from verified users in Sina Weibo. Experimental results show that the performance of our approach can achieve as high as 89.01%.

In future work, we will explore other latent attributions in social media such as age and job. A uniform system will be built to identify these attributes and explore their relationships.

References

1. Burger, J.D., Henderson, J., Kim, G., Zarrella, G.: Discriminating gender on twitter. In: Proceedings of the Conference on Empirical Methods in Natural Language Processing, pp. 1301–1309. Association for Computational Linguistics (2011)
2. Che, W., Li, Z., Liu, T.: Ltp: A Chinese language technology platform. In: Proceedings of the 23rd International Conference on Computational Linguistics: Demonstrations, pp. 13–16. Association for Computational Linguistics (2010)
3. Herring, S.C., Paolillo, J.C.: Gender and genre variation in weblogs. Journal of Sociolinguistics 10(4), 439–459 (2006)
4. Mukherjee, A., Liu, B.: Improving gender classification of blog authors. In: Proceedings of the 2010 Conference on Empirical Methods in Natural Language Processing, pp. 207–217. Association for Computational Linguistics (2010)
5. Nguyen, D., Smith, N.A., Rosé, C.P.: Author age prediction from text using linear regression. In: Proceedings of the 5th ACL-HLT Workshop on Language Technology for Cultural Heritage, Social Sciences, and Humanities, pp. 115–123. Association for Computational Linguistics (2011)
6. Peersman, C., Daelemans, W., Van Vaerenbergh, L.: Predicting age and gender in online social networks. In: Proceedings of the 3rd International Workshop on Search and Mining User-generated Contents, pp. 37–44. ACM (2011)
7. Rao, D., Yarowsky, D., Shreevats, A., Gupta, M.: Classifying latent user attributes in twitter. In: Proceedings of the 2nd International Workshop on Search and Mining User-generated Contents, pp. 37–44. ACM (2010)
8. Rosenthal, S., McKeown, K.: Age prediction in blogs: A study of style, content, and on-line behavior in pre-and post-social media generations. In: Proceedings of the 49th Annual Meeting of the Association for Computational Linguistics: Human Language Technologies, vol. 1, pp. 763–772. Association for Computational Linguistics (2011)
9. Salton, G., McGill, M.J.: Introduction to modern information retrieval (1983)

A Hybrid Method of Domain Lexicon Construction for Opinion Targets Extraction Using Syntax and Semantics

Chun Liao, Chong Feng, Sen Yang, and Heyan Huang

Department of Computer Science and Technology
Beijing Institute of Technology, Beijing 100081, China
{cliao,fengchong,syang,hhy63}@bit.edu.cn

Abstract. Considering opinion targets extraction of Chinese microblogs plays an important role in opinion mining, there has been a significant progress in this area recently, especially the CRF-based method. However, this method only takes lexical-related features into consideration and does not excavate the implied semantic and syntactic knowledge. We propose a new approach which incorporates domain lexicon with groups of features using syntax and semantics. The approach acquires domain lexicon through a novel way namely PDSP. And then we combine the domain lexicon with opinion targets extracted from CRF with groups of features together for opinion targets extraction. Experimental results on COAE2014 dataset show that this approach notably outperforms other baselines of opinion targets extraction.

Keywords: opinion targets extraction, domain lexicon, groups of features, CRF.

1 Introduction

With the prevalence of microblog in China, more and more people are using microblog for real-time communication and opinion diffusion. Therefore, opinion target extraction, as one subtask of opinion mining, is of great significance for both network monitoring and sentiment analysis.

Microblog is different from other media, the use of language in it has its own features[1,2]. For example microblog is usually short, interactive, non-standard and often contains many Out-of-Vocabulary words. Existing opinion targets extraction methods only take lexical-related features into consideration and do not excavate the implied syntactic and semantic knowledge. To address these shortcomings, it is intuitive to consider the combination of domain lexicon and groups of features using syntax and semantics, and at the same time take special features of microblog into consideration. In this paper we propose to acquire domain lexicon by a hybrid method of PDSP, and get groups of features through mining syntactic and semantic knowledge. Then, we regard opinion targets extraction as a sequence labeling task and choose different groups of features for CRF to finish the accurate extraction. In experiments on the

H.-Y. Huang et al. (Eds.): SMP 2014, CCIS 489, pp. 108–116, 2014.

COAE 2014 dataset we find that our method can substantially extract opinion targets more effectively under different evaluation metrics.

2   Related Work

The methods of opinion targets extraction on sentence level[3] are mainly divided into two categories: supervised and unsupervised methods. In the unsupervised methods, Hu and Liu[4] used association rules to excavate opinion targets and regarded the top-frequency words as opinion targets. Li and Zhou[5]extracted tuples like <emotional words, opinion targets> based on emotional and topic-related lexicons. Popescu and Nguyen[6] extracted properties of products with mutual information. Then with the convening of Chinese Opinion Analysis Evaluation conference in China, rule-based approaches were employed for opinion targets extraction. Liu[7] used syntactic analysis to obtain the candidates , and then combined PMI with noun pruning algorithm to decide the final opinion targets. Besides, in the supervised methods, Zhuang[8] proposed a multi-knowledge-based approach which integrated WordNet, statistical analysis and movie knowledge. Jakob[9]modeled the task as a sequence labeling question and employed CRF for opinion targets extraction. Until 2014, the 6th Chinese Opinion Analysis Evaluation proposed the task named elements extraction of opinion sentences in microblog. The task required to extract the opinion targets of one sentence in a collection of opinion sentences.

However, existing opinion targets extraction methods only took lexical-related features into account. Consequently, considering the specific features of Chinese microblog, we propose a new method for opinion targets extraction towards microblog using syntax and semantics in which we adopt a new approach of PDSP for domain lexicon construction and select groups of features for CRF.

3   PDSP: A Hybrid Method of Domain Lexicon Construction Using Syntactic and Semantics

The task of domain lexicon construction is automatically extracting the opinion targets using rule-based methods. Considering the fact that the quality of domain lexicon construction varies substantially with different kinds of extracting methods, we propose a new approach named PDSP which incorporates POS, dependency structure, semantic role and phrase structure for domain lexicon construction.

- Part of Speech
  Opinion targets are usually nouns or noun phrases. Through statistics on corpus, we design six templates based on Part-of-Speech which are shown in Table 1 where n, adj, adv, aw, cmp and OT represents for noun, adjective, degree adverb, advocating word, comparative word and opinion target. Here we get adv and aw from Hownet[1], and acquire cmp from [10].

---

[1] http://www.keenage.com/html/c_index.html

Table 1. Part-of-Speech sequence templates

Template	Example	Template	Example
n+adv+adj	屏幕/OT 很 好	adj+的+n	轻薄 的 机身/OT
n+adj	外观/OT 漂亮	n+cmp+n	iphone/OT不如 三星/OT
aw+n	认为 蒙牛/OT	n+n	蒙牛牛奶/OT

- Dependency Structure

  As we all know, when we express opinions towards a product, we need some opinion words which usually have strong semantic relation with the opinion targets. Therefore, we collect opinion words from Hownet and NTUSD [2] and perform HIT-LTP [3] for dependency parsing to discuss the relation "ATT" and "SBV" between opinion words and opinion targets, relation "COO" between already known opinion targets and unknown opinion targets. For example, "效率和画质都好于一般摄像头。", its dependency analysis result is illustrated in Fig.1. We can get "效率" as opinion target through opinion word "好" with the relation of "SBV" and acquire "画质" as opinion target through already known opinion target "效率" with the relation of "COO".

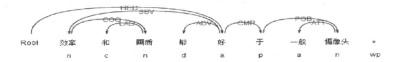

Fig. 1. Example of dependency analysis result

- Semantic Role

  As a necessary part of shallow semantic parsing, sematic role[11] occupies an important position in lexical and semantic analysis. People usually express opinions through opinion words in opinion sentences. And adjective and verb are two main forms of opinion words. Through investigation, we find when the opinion word are adjective, A0(agent) is opinion target. Furthermore, when the opinion word are verb, A1(patient) is opinion target. The example of semantic role labeling result is illustrated in Fig.2.

尼康	D7000	的	外观	很	漂亮		我	喜欢	尼康	D7000
nz	ws	u	n	d	a		r	v	nz	ws

| | A0 | | ADV | 漂亮 | | A0 | 喜欢 | A1 |
(a)   (b)

Fig. 2. Example of semantic role labeling result

Finally, we analyze the semantic role labeling result and summarize the opinion targets extraction algorithms based on semantic roles as follows:

Algorithm 1: opinion targets extraction algorithm

```
Input: tokenized sentences, POS, emotional words,
semantic role labeling results SRL
Output: opinion targets
for word in sentences:
  if word in emotional words:
    if word.POS = adj:
      return word in the sentence which SRL=A0
    if word.POS = v:
      return word in the sentence which SRL=A1
```

- Phrase Structure

  For example, we perform phrase structure analysis on "好喜欢红色的奥迪A4, 比奔驰C200好很多。" using Stanford Parser, its phrase structure tree is illustrated in Fig.3. As we can see in the tree, "奥迪","A4" are not regarded as a unique node of "奥迪A4", so that the opinion targets we finally extracted are "奥迪"and"A4", not "奥迪A4". Considering this, to improve the boundary identification ability of opinion targets, we merge them with their labels of parent nodes through statistics on corpus, such as "奥迪""A4"can be combined by the label "NR+CD".

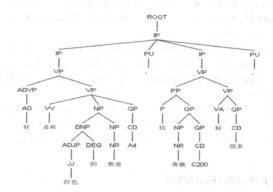

Fig. 3. Example of phrase structure analysis result

Consequently, the algorithm PDSP is summarized as following 5 steps:

- Use the six POS sequence templates in Table 1 to match each sentence in corpus and select the nouns or noun phrases in the matched sentences as the opinion targets. We regard these opinion targets as set A.
- Employ dependency parsing and extract opinion targets as set B through analyzing the relation of SBV, ATT and COO.

- Conduct the opinion targets extraction algorithm of Algorithm 1 based on semantic roles to extract opinion targets as set C.
- Compute set D=A+B+C.
- Traverse set D and combine the opinion targets based on phrase structure to finally complete the domain lexicon construction.

4   Selection on Groups of Features with Lexical Semantics and Syntactic Dependencies

In CRF-based method, the features we employed as input are of great importance. In this section, we refer to the features which are employed by Jakob[9] and Lu[12] in English and meanwhile put forward some new features based on the specific grammar of Chinese. Generally, we think opinion targets extraction is primarily related with four kinds of features which are named as lexical features, dependency features, relative position features and semantic features.

- Lexical Features
  As words with the same Part-of-Speech usually appear around the opinion targets, we select the current word itself and the POS of current word as lexical features.
- Dependency Features
  Dependency parsing reflects the semantic dependency relations between core word and its subsidiaries words[13]. Consequently, we select whether the dependency between current word and core word exists, the dependency type, parent word and the POS of parent word as the dependency features.
- Relative Position Features
  As we all know, since words which appear around emotional words are more likely to be opinion targets, we determine the boolean value by judging whether the distance between current word and emotional word is less than 5.
- Semantic Features
  As stated in section 3, there is a strong relationship between the sematic roles and POS of emotional words. Considering this, we select the sematic role name of current word and POS of emotional word in this sentence for CRF.

5   Experiments and Analysis

In experiments, we firstly obtained opinion targets named set X through the method of domain lexicon construction in section 3, and then we employed CRF with the domain lexicon and features in section 4 together to extract opinion targets namely set Y, finally we combine set X and Y as the final extracted opinion targets.

5.1   Preparation of Dataset

Considering the short, interactive, non-standard features of microblog, we design some rules to normalize the corpus.

Rule 1: Delete the English and interrogative sentences;

Rule 2: Turn over the sentences with "//" to ensure the forwarding relationship;

Rule 3: Delete structures like "@+用户名", "http://t.cn/h87oy";

Rule 4: Replace the consecutive punctuations with the first one and substitute the Chinese explanation for emoticon;

Rule 5: For the special structure "#content#", if the length of the content is short, we choose it as candidate opinion targets. And if not, as an independent clause;

Rule 6: Perform anaphora resolution for the sentences with pronouns.

Rule 7: Delete sentences which do not contain emotional words.

Through filtration by these rules, we finally obtain 5,000 normalized sentences with opinion orientation. In this paper, we conduct experiments on such a dataset and assess it with traditional Precision, Recall and F-measure under strict and lenient evaluations which respectively represents the extraction result is exactly the same or overlapped with the labeled one.

5.2   Comparison between Different Methods of Domain lexicon Construction

As the quality of candidate domain lexicon greatly influences the result of opinion targets extraction, we make a comparison of different approaches of domain lexicon construction in this section. We conduct experiments of domain lexicon construction with methods of POS [14], syntactic [15], semantic role and PDSP, and the comparing results of these four approaches are represented in Table 2.

Table 2. Comparing results of different domain lexicon construction methods

Method	Strict Evaluation			Lenient evaluation		
	Precision	Recall	F-measure	Precision	Recall	F-measure
POS[14]	0.3238	0.4196	0.3655	0. 3436	0.4424	0.3867
Syntactic[15]	0.3452	0.4388	0.3864	0. 3750	0.4492	0.4087
Sematic Role	0.3362	0.4261	0.3758	0. 3621	0.4518	0.4020
PDSP	**0.3821**	**0.4834**	**0.4268**	**0. 4122**	**0.4943**	**0.4495**

The results show that the method PDSP proposed by this paper performs notably better than the other ones. The effectiveness of PDSP is primarily because its combination of four kinds of important information: POS, dependency, phrase structure and semantic role information.

5.3   Analysis on Groups of Features

In this section, we adopt four kinds of different groups of features for CRF. To verify the effectiveness of features proposed in section 4, we conduct experiments under different groups of features and obtain the experimental results as shown in Table 3. And the lexi, pos, dp, srl and dl respectively represent for lexical, relative position, dependency, semantic and domain lexicon feature.

Table 3. Results of CRF with different groups of features

Method	Strict Evaluation			Lenient evaluation		
	Percision	Recall	F-measure	Precision	Recall	F-measure
lex	0.6470	0.4106	0.5023	0.6932	0.4272	0.5286
lex+pos	0.6538	0.4196	0.5111	0.7040	0.4316	0.5351
lex+pos+dp	0.6673	0.4261	0.5200	0.7134	0.4463	0.5490
lex+pos+dp+srl	0.6824	0.4472	0.5403	0.7374	0.4685	0.5729
lex+pos+dp+srl+dl	**0.6864**	**0.4513**	**0.5446**	**0.7398**	**0.4712**	**0.5757**

It can be seen that the effect of opinion targets extraction is highly improved after adding dependency, semantic and domain lexicon features, which is probably because they can explore the implied syntactic and semantic information. For example, "好想要啊，三星手机的音质非常赞！", traditional method can just extract "三星手机" as the opinion target. However, when the semantic feature is added into CRF, we can certainly extract "三星手机的音质" as the opinion target.

5.4   Combination of Domain Lexicon and Groups of Features

We compare the combination of domain lexicon and groups of features method namely 4 in Fig.4 with the other three basic methods as follows: lexicon-based, Jakob[9] and the best of CRF-based method in section 5.3 which are respectively represented as 1,2,3 in Fig.4.

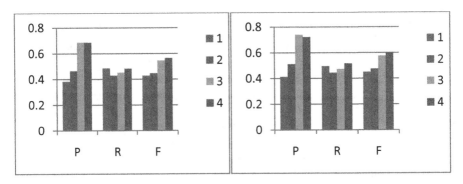

Fig. 4. Experimental results under strict evaluation and lenient evaluation

As we can see, the approach of integrating domain lexicon with groups of features improves the result substantially which is mainly because this method not only uses domain lexicon constructed in section 3 to obtain a part of opinion targets, but also adopts machine learning method of CRF to make up for the defect of rule-based method and so as to reach a higher precision, recall and F-measure. So this experiment strongly demonstrates the effectiveness and applicability of combination of domain lexicon and groups of features method.

6   Conclusions and Future Work

In this paper we propose a hybrid method of domain lexicon construction which takes Part-of-Speech, dependency structure, semantic role and phrase structure into consideration. We combine the domain lexicon with the opinion targets extracted from CRF with lexical, dependency, position and semantic features to get the final opinion targets. The experimental results show that it performs better than other baseline approaches with different domain lexicon construction methods and groups of features.

In the future work, we will take the following points into consideration:

- Considering the various expressions of Chinese microblog, we should excavate more rules and extract the kernel sentence for opinion targets extraction.
- In the feature selection for CRF, we will make further research on it and excavate more useful features to improve the performance of CRF.
- In this paper, we perform opinion targets extraction on sentence level. We will investigate the effect of opinion targets extraction on corpus level.

Acknowledgements. The work described in this paper was supported by the National Basic Research Pro-gram of China (973 Program, Grant No. 2013CB329605, 2013CB329303) and Na-tional Natural Science Foundation of China (Grant No. 61201351). We would like to thank COAE[16] for offering this dataset. We would also acknowledge the help of HIT-IR-Lab for providing the Chinese dependency parser[17] and Stanford for offering the Stanford Parser[18].

References

1. Liu, Q., Feng, C., Huang, H.: Emotional Tendency Identification for Micro-blog Topics Based on Multiple Characteristics. In: 26th Pacific Asia Conference on Language, Information and Computation, pp. 280–288 (2012)
2. Han, Z., Zhang, Y., Zhang, H., Wan, Y., Huang, J.: An effective short text tendency classification algorithm for Chinese microblogging. Computer Application and Software 29(10), 89–93 (2012)
3. Rao, D., Ravichandran, D.: Semi-supervised polarity lexicon induction. In: Proceedings of the 12th Conference of the European Chapter of the Association for Computational Linguistics, pp. 675–682 (March 2009)
4. Hu, M., Liu, B.: Mining and summarizing customer reviews. In: Proceedings of the Tenth ACM SIGKDD International Conference on Knowledge Discovery and Data Mining, pp. 168–177 (August 2004)
5. Li, B., Zhou, L., Feng, S., Wong, K.F.: A unified graph model for sentence-based opinion retrieval. In: Proceedings of the 48th Annual Meeting of the Association for Computational Linguistics, pages, pp. 1367–1375 (July 2010)
6. Popescu, A.M., Etzioni, O.: Extracting product features and opinions from reviews. In: Natural language Processing and Text Mining, pp. 9–28 (2007)
7. Liu, H., Zhao, Y., Qin, B., Liu, T.: Comment Target Extraction and Sentiment Classification. Journal of Chinese Information Processing 24(1), 84–88 (2010)

8. Zhuang, L., Jing, F., Zhu, X.Y.: Movie review mining and summarization. In: Proceedings of the 15th ACM International Conference on Information and knowledge Management, pp. 43–50 (November 2006)

9. Jakob, N., Gurevych, I.: Extracting opinion targets in a single-and cross-domain setting with conditional random fields. In: Proceedings of the 2010 Conference on Empirical Methods in Natural Language Processing, pp. 1035–1045 (October 2010)

10. Zhang, C., Feng, C., Liu, Q., Shi, C., Huang, H., Zhou, H.: Chinese Comparative Sentence Identification Based on Multi-feature Fusion. Journal of Chinese Information Processing 27(6), 110–116 (2013)

11. Hacioglu, K.: Semantic role labeling using dependency trees. Computational Linguistics, 1273 (August 2004)

12. Lu, B.: Identifying opinion holders and targets with dependency parser in Chinese news texts. In: Proceedings of the NAACL HLT 2010 Student Research Workshop, pp. 46–51 (June 2010)

13. Li, X., Roth, D.: Learning question classifiers. In: Proceedings of the 19th International Conference on Computational Linguistics, pp. 1–7 (August 2002)

14. Zhu, Y., Xu, Y., Wang, W., Lu, L., Du, R., Deng, C.: Research on Opinion Extraction of Chinese Review. In: Proceedings of the Third Chinese Opinion Analysis Evaluation, pp. 126–135 (2011)

15. Sun, H., Yu, S.: Shallow Parsing: An overview. In: Contemporary Linguistics, pp. 74–83 (2000)

16. http://www.liip.cn/CCIR2014/pc.html

17. Che, W., Li, Z., Liu, T.: Ltp: A chinese language technology platform. In: Proceedings of the 23rd International Conference on Computational Linguistics: Demonstrations, pp. 13–16 (August 2010)

18. Socher, R., Bauer, J., Manning, C.D., Ng, A.Y.: Parsing with compositional vector grammars. In: Proceedings of the 51st Annual Meeting of the Association for Computational Linguistics, pp. 455–465 (2013)

Online Social Network Profile Linkage
Based on Cost-Sensitive Feature Acquisition

Haochen Zhang[1], Minyen Kan[2], Yiqun Liu[1], and Shaoping Ma[1,*]

[1] State Key Laboratory of Intelligent Technology and Systems,
Tsinghua National Laboratory for Information Science and Technology,
Department of Computer Science and Technology,
Tsinghua University, Beijing 100084, China
[2] Web, IR / NLP Group (WING)
Department of Computer Science, National University of Singapore, Singapore
zhang-hc10@mails.tsinghua.edu.cn
kanmy@comp.nus.edu.sg
{yiqunliu,msp}@tsinghua.edu.cn

Abstract. Billions of people spend their virtual life time on hundreds of social networking sites for different social needs. Each social footprint of a person in a particular social networking site reflects some special aspects of himself. To adequately investigate a user's preference for applications such as recommendation and executive search, we need to connect up all these aspects to generate a comprehensive profile of the identity. Profile linkage provides an effective solution to identify the same identity's profiles from different social networks.

With various types of resources, comparing profiles may require plenty of expensive and time-consuming features such as avatars. To boost the online social network profile linkage solution, we propose a cost-sensitive approach that only acquires these expensive and time-consuming features when needed. By evaluating on the real-world datasets from Twitter and LinkedIn, our approach performs at over 85% F_1-measure and has the ability to prune over 80% of the unnecessary feature acquisitions, at a marginal cost of 10% performance loss.

Keywords: social media, user profiles, profile linkage, cost-sensitive.

1 Introduction

Online social network is the most important part of the cyber-life, where netizens share their lives, express their opinions, communicate with their friends and business partners. People use more than one social network to satisfy different social needs of sharing, reading, discussing and communicating. He may

* This work was supported by Natural Science Foundation of China (61073071). Part of the work was done at the Tsinghua-NUS NExT Search Centre, which is supported by the Singapore National Research Foundation & Interactive Digital Media R&D Program Office, under research grant R-252-300-001-490.

H.-Y. Huang et al. (Eds.): SMP 2014, CCIS 489, pp. 117–128, 2014.
© Springer-Verlag Berlin Heidelberg 2014

communicate his friends in Facebook, post his comments in Twitter, show his life in Instagram and connect to his business in LinkedIn. To picture a person completely, especially for executive search and recommendation, it is very important to cover all aspects of the person's virtual footprints. Therefore, finding an effective solution to identify users for the same identity has high attractiveness in academic study and commercial value in business.

The similar task, record linkage, has been investigated in traditional database research area for decades. There is also several related work addressing the problem in social network perspective recently[1–7]. However, these approaches barely apply to the large-scale dataset and fail to consider the difficulty dealing with the time-consuming and expensive feature acquisitions. In this paper, we propose an effective and efficient approach taking both features directly extracted from profiles and expensive features acquiring cost-sensitively.

The remainder of this paper is structured as follows. We first describe the related work that informs our task in the next section. In Section 3, we define our problem and describe our analysis of online social network user profiles. This motivates our chosen method to maximally leverage well-populated attributes in profiles for profile linkage, which we present in Section 4. In Section 5, we evaluate our approach on it to examine both effectiveness and efficiency.

2   Related Work

2.1   Profile Linkage across Social Networks

Although profile linkage problem just rises along with the booming development of online social networks, the related task record linkage, also named as entity resolution, has been well studied in traditional database area, including named attributes computations[8, 9], schema mapping for heterogeneous data structures[10–12], probabilistic linkage models[13] and duplicate detection for hierarchical-structured data[14].

Inherited from record linkage task, several work addresses the profile linkage task by applying the intuitive attribute comparison approaches into social network occasions[1–4, 15]. Liu *et al.* [16] and Zafarani *et al.* [7] carefully investigate behaviors of how a user generate his username, and then discover user's characteristics to identify the same individual. Besides attributes comparison, Narayanan *et al.* [5] and Bartunov *et al.* [6] rely on social connections and settle identification by exploiting the assumption that a person has similar social circles across different web sites.

However, these work is not based on the real world dataset, which ignores the problems of time-consuming and expensive feature acquisition procedures. When dealing with large-scale data, the enormous cost has to be considered.

2.2   Cost Sensitive Feature Acquisition

Traditional linkage tasks usually gather all attributes locally and features are easily generated. Thus the cost of acquiring and computing features is omitted.

However, both employing web services results acquiring external resources from web and in extremely high cost comparing to local similarity computation.

Several approaches are investigated to deal with missing attribute values acquisition [17–19]. Lin *et al.* [20] improved probabilistic K-NN with acquiring attribute values of uncertain data objects. Tan *et al.* [21] proposes a hierarchical cost-sensitive approach to acquire search engine results with hierarchical dependencies. However, user profile linkage task acquires various types of features with both time cost and usage limitations. These approaches are not designed to solve the profile linkage tasks in which we should consider the hybrid cost controlling.

3   Motivation

3.1   Profile Linkage

Identity refers to a unique entity, such as individual people, groups and companies, which is usually identifiable in the real-world. **Profile** refers to a particular social network's account for the identity, which consists of **attributes** with values. In different social networks, an identity may register several accounts to cover different social applications. Intuitively, profiles from the same identity should be quite similar to each other. **Profile linkage** is then defined as the task that discovers profiles projected from the same identity. Similar to other linkage tasks, profile linkage has two kinds of solutions: 1) clustering profiles for a certain identity; 2) comparing each pair of profiles to determine whether they belong to the same identity.

We address the profile linkage problem by comparing candidate profile pairs, denoted as *pairwise profile linkage*. Notice that there exist transitivity conflicts when involving more than two social networks. In this paper, we only solve the case of two social networks and leave the transitivity conflicts in future work.

The online social network profile linkage faces unique characteristics, including semi-structured data, multimedia resources, privacy and so on. Therefore, attributes in profiles are often sparse and arbitrary. Meanwhile, profiles in different OSN often prefer different attributes. As an example, Twitter is most public and similar to most other closed social media (i.e., FaceBook), where people share their personality to attract followers.

To take full advantage of online social network profiles, we adopt two external features to solve the problem of sparse and arbitrary features: 1) Geocode is a kind of structured locality information, which is much more precise than comparing textual location. Google Maps API provides web service to convert strings to geocodes. 2) Avatar is the most common multimedia resources in profile. A person may use same portrait across different social networks, which provides a very strong evidence when distinguishing between people with the same name.

3.2   Cost-Sensitivity

Since adopting time-consuming features such as geocode and avatar, we face a trade-off between the feature acquisition cost and the classifier performance.

The acquisition of the complicated features could be very time-consuming. The web services even have limitations or payments for the usage. The expensive feature acquisitions motivate the cost-sensitive approach that carefully selects the most distinguishable features with less cost. We regard the selective feature acquisitions that reduce the time and network cost as the micro-level cost-sensitivity.

In this work, the object of the approach is to achieve better results with less time-consuming within the given usage quotas. To every comparison having available extra feature, we define that the comparison benefits if it is re-classified to be correct by given certain extra feature. Therefore, we predict the expected benefit of a comparison in unit time-consuming and adopt the expectation as the criteria of selecting the most effective feature.

4   Approach

To solve the online social network profile linkage, we propose an indexing framework. We index all profiles by tokens extracted from usernames. The tokens are the continuous letters or digit sequences separated by spaces or symbols. Based on Liu et al. [16]'s survey, 79% users prefer same username across different communities. Our investigation also shows that 96.1% matched profiles are connected by at least one token. Therefore, two profiles are very unlikely to be matched without a shared token, which helps to prune unnecessary pair-wise comparisons.

Afterwards, we retrieve all pairs of profiles that share at least one token and adopt a probabilistic classifier to determine whether the given two profiles are from the same identity.

4.1   Probabilistic Classifier

The probabilistic classifier is employed to estimate the probability of whether two given profiles q and t are linked (denoted as $l_{q,t} = \{0, 1\}$), by given similarity features $F_{q,t}$ and shared tokens $M_{q,t}$. By assuming the similarity features and shared tokens are independent of each other, we have:

$$p_{q,t} = Pr(l_{q,t}|F_{q,t}, M_{q,t}) = \frac{Pr(l_{q,t}|M_{q,t}) \times \prod_{f_k \in F_{q,t}} Pr(f_k|l_{q,t})}{\prod_{f_k \in F_{q,t}} Pr(f_k)} \tag{1}$$

where $Pr(l_{q,t}|M_{q,t})$ is approximately calculated by:

$$\hat{Pr}(l_{q,t} = 1|M_{q,t}) = \frac{1}{|\bigcap_{m \in M_{q,t}} D_m| + \beta} \tag{2}$$

where D_m is all profiles indexed by token m and β is a smoothing factor preventing $Pr(l_{q,t}|M_{q,t})$ from being 1. We set $\beta = 0.5$ empirically in our experiments.

By applying $Pr(l_{q,t} = 0|\cdot) + Pr(l_{q,t} = 1|\cdot) = 1$ to Equation 1, the equation is derived:

$$p_{q,t} = \frac{1}{1 + (|\bigcap_{m \in M_{q,t}} D_m| + \beta - 1) \times \prod_{f_k \in F_{q,t}} \frac{Pr(f_k|l_{q,t} = 0)}{Pr(f_k|l_{q,t} = 1)}} \tag{3}$$

where $\hat{Pr}(f_k|l_{q,t})$ is estimated by kernel density estimator [22, 23]. Finally, q and t are regarded as matched if $p_{q,t} > 0.5$.

4.2   Features

CSPL uses features discovered from user's profile to conduct the supervised linkage. These features are consist of local features extracted directly from profile attributes, and external features acquired by web services or web resources. To estimate the benefit for cost-sensitive acquisition, all features are normalized to a range of $[0, 1]$. Table 1 lists all the involved local features and how they are computed.

Table 1. Local similarity features extracted directly from profiles

Feature	Description
user_sim	Jaro Winkler distance between two usernames
language	Jaccard similarity of the written or spoken languages
description	vector-space model cosine similarity of user's biography
URL	cosine similarity of the URL tokens (split by symbols)
popularity	$\frac{|friend_q - friend_t|}{|friend_q + friend_t|}$ where $friend_u$ is the counts of user u's friends

Besides these easily acquired local features, we also employ two time-consuming and usage-limited features: **avatar** and **geocode** as discussed in section 3:

1. **Avatar** is user uploaded image, given as a URL in the profile. We employ a gray-scale χ^2 dissimilarity, a bin-by-bin histogram difference by [24], to compare avatars. This method has been reported effective for texture classification, and represented as:

$$F_{avatar} = \frac{1}{2} \sum_{i \in Bins} \frac{(H_{q,i} - H_{t,i})^2}{(H_{q,i} + H_{t,i})} \tag{4}$$

   where $H_{q,i}$ and $H_{t,i}$ is the ith bin of the image's gray-scale histograms.
2. **Geocode** is the structured information with lat-long coordinates. We access Google Maps API to convert textual location into geographic coordinates, and then calculate spherical distance d in kilometers for comparison. At last, we use $e^{-\gamma d}$ to normalize the distance within $[0, 1]$ with $\gamma = 0.001$ in our experiments.

4.3   Cost-Sensitive Feature Acquisition

Note that some of our features are externally acquired. For example, obtaining the Geocode requires API invocations, while obtaining a users' Avatar requires a separate resource request. These external features are expensive to acquire as they incur both network delays and bandwidth, and are much more costly than computation over local features. We wish to manage these costs, so as not to use external resources when the local evidence already overwhelmingly supports a linkage decision.

In fact, compared to to acquiring external features, the cost of computation over all local features is negligible. In our classifier, we thus first utilize all of the local features in the beginning, and iteratively choose the instances that are most probable to be improved by adopting a new external feature. We employ a cost-sensitive classifier derived from Naïve Bayes to prune these unnecessary network operations.

Let $\hat{p}$ denote the probability distribution estimated by the existing set of features, and the $\hat{p}^{+k'}$ be the posterior probability when conditioned on the additional feature k'. Let $f_{k'}$ be value of the extra feature, and we derive the relationship between $\hat{p}$ and $\hat{p}^{+k'}$:

$$\hat{p}^{+k'} = \frac{1}{1 + \dfrac{1-\hat{p}}{\hat{p}} \times \dfrac{Pr(f_{k'}|l=0)}{Pr(f_{k'}|l=1)}} \tag{5}$$

To efficiently improve linkage performance, we need to acquire the k' feature that is most effective. Here, we assume that adding features improves performance at the entire dataset level. We estimate the benefit of acquiring a prospective feature by its utility to raise the certainty of the classifications, either for a match or a non-match:

$$(\hat{p} - 0.5)(\hat{p}^{+k'} - 0.5) \leq 0 \tag{6}$$

By solving the inequality, we can restate the post-condition of the classification:

$$\begin{cases} g_{k'} < \hat{p}^{-1} - 1, & \hat{p} > 0.5 \\ g_{k'} > \hat{p}^{-1} - 1, & \hat{p} \leq 0.5 \end{cases} \tag{7}$$

where $g_{k'}$ represents the ratio $\dfrac{\hat{Pr}(f_{k'}|l=1)}{\hat{Pr}(f_{k'}|l=0)}$ for convenience.

Note that $f_{k'}$ is unknown and cannot be computed directly ahead of acquisition. We therefore must estimate the probability that $g_{k'}$ satisfies the condition, given $\hat{p}$, which is difficult to compute accurately as the distribution of $g_{k'}$ is unknown. We thus need to develop an approximation method.

Notice that we have already estimated $Pr(f_{k'}|l)$ by using the kernel density estimator during training. Furthermore, all our features have values in normalized range of $[0,1]$. We sample s points $\Delta_1, \Delta_2, \cdots, \Delta_s$ within $[0,1]$ following the distribution of $Pr(f_{k'})$ estimated during training using the kernel density estimation. We then compute the corresponding estimation $\hat{g}_{k'j}|_{f'=\Delta_j}$. Suppose that $r_{p,k'}$ is the rank of value $\hat{p}^{-1} - 1$ in the ordered list of $\{\hat{g}_{k'}\}$, we can then compute the approximate benefit of acquiring feature k' with equation:

$$E_{q,t}^{k'} = \hat{Pr}(\text{benefit}|p, k') = \begin{cases} \dfrac{r_{p,k'}}{s}, & p > 0.5 \\ 1 - \dfrac{r_{p,k'}}{s}, & p \leq 0.5 \end{cases} \tag{8}$$

where $E_{q,t}^{k'}$ is the expectation of benefit given $\hat{p}$ and k'.

In practice, acquiring different features has different time costs. This adds another dimension of complexity to our feature acquisition process, as features have different per time-unit benefit. If the time cost of acquiring feature k' is $c_{k'}$, the *per time-unit feature benefit expectation* of comparison with probability p and added feature k' is $\dfrac{Pr(\text{benefit}|p, k')}{c_{k'}}$.

We conclude the *per time-unit benefit expectation* of the given comparison $\langle q, t \rangle$ to be:

$$E_{q,t} = \max_{k' \in \mathcal{K}_{q,t}} \frac{E_{p_{q,t}}^{+k'}}{c_{k'}} \tag{9}$$

where $\mathcal{K}_{q,t}$ represents the set of acquirable features of the comparison $\langle q, t \rangle$. By acquirable, we mean that a feature meets the following criteria: 1) have not been acquired; 2) exist in both profiles; 3) its acquisition will not exceed a quota (e.g., a API daily limit). The most effective external feature is the one that maximizes the benefit expectation.

5   Experiment

We set up experiments on linking 150,000 users across Twitter and LinkedIn to evaluate the performance of our linkage approach and the efficiency and effectiveness of the cost-sensitive feature acquisition method.

5.1   Linkage Performance

To evaluate the classifier's performance, we crawled a realistic profile dataset from Twitter and LinkedIn. The Twitter profiles are sampled from tweets posted between 9 Oct. 2012 and 16 Oct. 2012. The LinkedIn profiles are sampled from the directory[1]. In total, we obtained 152,294 Twitter profiles by RESTful API and 154,379 LinkedIn profiles by parsing user profile pages, which are all publicly available.

To discover the relationship between each LinkedIn and Twitter profiles, we employ third party websites Google+ that encourages users to reveal their OSN profiles. We generate the ground truth with the assumption that all corresponding OSN accounts filled by one user belong to themselves. We crawled 180,000 Google+ profiles and extract the overlapping users of our dataset and the Google+ profiles. This partial ground truth contains 9,750 identities: 4,779 matched Twitter–LinkedIn users, 3,339 singular Twitter users and 1,632 singular LinkedIn users.

Besides the standard IR metrics: Precision (Pre), Recall (Rec) and F_1-measure (F_1), we employ the identity-based accuracy (I-Acc), representing as:

$$I\text{-}Acc = \frac{\text{correctly identified identities}}{\text{all ground truth identities}}$$

---

[1] http://www.linkedin.com/directory/people/a.html

We use simple classifiers like C4.5, SVM and Naïve Bayes as the base-line, which have been reported effective in [2]. In our experiment, we choose Twitter as the query dataset and LinkedIn as the target. To generate the training set, We randomly sampled 1,000 query instances and all the corresponding targets. To evaluate the classifier performance adequately, we include all features in this experiment. Table 2 shows that our approach CSPL make the best performance both in F_1 and I-Acc. Furthermore, our approach make a significant improvement in recall with slight loss in precision, which discovers more linking relationship between candidate pairs.

Table 2. Linkage performance over our Twitter→LinkedIn dataset with all features

Method	Pre	Rec	F_1	I-Acc
C4.5	0.905	0.658	0.762	0.806
SVM	0.942	0.456	0.614	0.727
Naïve Bayes	0.934	0.625	0.748	0.801
CSPL	0.866	**0.846**	**0.856**	**0.865**

5.2 Cost-Sensitive Feature Acquisition

CSPL is also designed to optimally control for cost in acquiring external features. We denote our cost-sensitive approach based on benefit expectation as described in Section 4.3 as CSPL_BE.

Table 3. Cost-sensitivity with different pseudo time cost settings

#I	$C_l = C_a = 1$ Acq	Time	$C_l = 1, C_a = 3$ Acq	Time	$C_l = 3, C_a = 1$ Acq	Time
30%	2,000	2,000	2,000	2,110	4,000	8,273
60%	5,000	5,000	5,000	6,224	6,000	13,202
90%	32,000	32,000	32,000	79,072	32,000	45,824
	Time (Max Acq = 188,590)					
100%	188,590		439,880		314,480	

Since CSPL_BE is based on the time-unit benefit expectation, it is sensitive to different time cost settings. To comprehensively evaluate the performance in different time cost settings, we set up and investigate three pseudo time settings to simulate possible cases: $\langle C_l, C_a \rangle = \langle 1, 1 \rangle$, $\langle C_l, C_a \rangle = \langle 1, 3 \rangle$ and $\langle C_l, C_a \rangle = \langle 3, 1 \rangle$, where C_l and C_a is the time cost of *Geocode* and *Avatar* respectively. The experiment results are sampled per 2,000 acquisitions. To make the results comparable, we set three checkpoints to estimate the approximate number of acquisitions and time cost to the nearest sample. Table 3 gives pseudo time costs over three checkpoints at 30%, 60% and 90% of all external feature acquisitions. Coupled with the results from Figure 1, we see that CSPL_BE achieved 90% of the remaining performance improvement that would be achieved by acquiring all external features, by merely acquiring 17% additional features and between 15–18% more time (depending on cost settings).

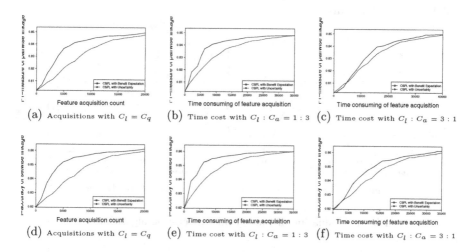

Fig. 1. Performance with different feature acquisition cost by given pseudo time cost. Figures in the first row are evaluated by F_1-measure and ones in the second row are evaluated by I-accuracy.

To specifically evaluate our benefit estimation, we need to compare with other forms to estimating the utility of yet unacquired features. We establish a baseline (CSPL_UN) that acquires the minimal cost feature of the most uncertain comparisons. The uncertainty of a comparison is defined as $1 - 2 \times |\hat{p} - 0.5|$, where $\hat{p}$ is the previous estimated probability of the comparison. Figure 1 displays this comparison for the three different pseudo time cost settings and two evaluation metrics. Note that acquiring only about 15% of the external is effective enough. Thus we only investigate the subfigure in which the performance has not reached stable status yet. Illustrated by the figures, our approach represented in solid line increases faster than the baseline represented in dashed line despite different time cost settings, which indicates that our approach performs better with the same feature acquisitions or time cost.

5.3   Benefit Expectation

Our cost-sensitive feature acquisition strategy relies on the estimation to the benefit expectation. We conduct this experiment to evaluate whether our algorithm is able to correctly simulate the expectation before actual feature acquisition.

We evenly split $[0, 1]$ into N buckets and assign the midpoint as the representative probability $\tilde{p}_i$ to the i-th bucket. We first initial the matched probability p of each comparison computed by local features. Each comparison is then dispatched into a certain bucket according to p. We acquire every external feature individually for these comparisons and get updated results. The percentage of beneficial instances within each bucket is then regarded as the actual benefit expectation to the representative probability $\tilde{p}_i$. Meanwhile, we have the set of probabilities $\tilde{p}_i$ and then compute their corresponding $E_p^{k'}$ without actually involving the new features.

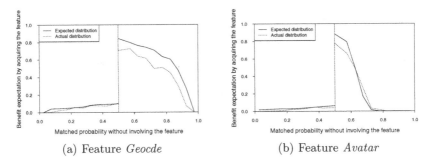

(a) Feature *Geocde*   (b) Feature *Avatar*

Fig. 2. Evaluation of the benefit expectation algorithm

In Figure 2, solid lines represent the prediction of the beneficial comparisons calculated by our algorithm, while dashed lines represent the actual expectation observed by actually involving new features. Figure 2 ((a)) and ((b)) corresponding to feature *Geocode* and *Avatar* respectively. Figure 2 illustrates that estimation and actual results are consistent, which indicates that our algorithm correctly estimates the benefit expectation.

In addition, we observe that the instances classified as positives, i.e. potential false positives, are more likely to be corrected than false negatives. It is mainly caused by the unbalanced dataset where there are too many negative instances. Meanwhile, the *Geocode* has a better benefit expectation with the same probability comparing to it Avatar, which indicates that *Geocode* is more effective in improving performance. It provides an intuitive way to compare the effectiveness between different features.

6   Conclusion

We investigate the profile linkage problem and propose a cost-sensitive probabilistic approach to reduce time consuming feature acquisitions. To effectively acquire external features, we establish an approximate algorithm to estimate the benefit of involving a new feature with performance time unit cost. The strategy is also able to satisfy the limitation of feature quota.

Our experiment results show the effectiveness of our approach with 85% F_1-measure and 86% I-accuracy compared to base-line. Our cost-sensitive framework also has the ability to prune the unnecessary network acquisitions for external features while keeping the performance loss in an acceptable level. Indeed, with only 10% loss on F_1-measure, we achieve more than 85% network acquisitions reduction.

In the future work, there are two routines to improve the linkage approach: 1. Improve linkage approach with more than profile resources, such as involving social connections, user generated content and mobile footprints. 2. Adopt the linked identities to a practical application such as recommendation to investigate the improvement caused by adequate user study.

References

1. Carmagnola, F., Cena, F.: User Identification for Cross-system Personalisation. Inf. Sci. 179(1-2) (2009)
2. Malhotra, A., Totti, L., Meira Jr., W., Kumaraguru, P., Almeida, V.: Studying User Footprints in Different Online Social Networks. In: International Workshop on Cybersecurity of Online Social Network (2012)
3. Nunes, A., Calado, P., Martins, B.: Resolving User Identities over Social Networks through Supervised Learning and Rich Similarity Features. In: Proceedings of the 27th Annual ACM Symposium on Applied Computing. ACM (2012)
4. Vosecky, J., Hong, D., Shen, V.: User Identification Across Multiple Social Networks. In: Networked Digital Technologies. IEEE (2009)
5. Narayanan, A., Shmatikov, V.: De-anonymizing Social Networks. In: Proceedings of the 2009 30th IEEE Symposium on Security and Privacy. IEEE (2009)
6. Bartunov, S., Korshunov, A., Park, S., Ryu, W., Lee, H.: Joint Link-Attribute User Identity Resolution in Online Social Networks. In: Proceedings of the 6th International Conference on Knowledge Discovery and Data Mining, Workshop on Social Network Mining and Analysis. ACM (2012)
7. Zafarani, R., Liu, H.: Connecting Users across Social Media Sites: A Behavioral-modeling Approach. In: Proceedings of the 19th ACM SIGKDD International Conference on Knowledge Discovery and Data Mining, pp. 41–49. ACM, New York (2013)
8. Cohen, W.W., Ravikumar, P., Fienberg, S.E., et al.: A Comparison of String Distance Metrics for Name-matching Tasks. In: Proceedings of the IJCAI 2003 Workshop on Information Integration on the Web (IIWeb 2003), pp. 73–78 (2003)
9. Christen, P.: A Comparison of Personal Name Matching: Techniques and Practical Issues. In: Proceedings of the 6th IEEE International Conference on Data Mining Workshops, ICDM Workshops. IEEE (2006)
10. Aumueller, D., Do, H.H., Massmann, S., Rahm, E.: Schema and Ontology Matching with Coma++. In: Proceedings of the 2005 ACM SIGMOD International Conference on Management of Data, SIGMOD 2005, p. 906. ACM Press (2005)
11. Nottelmann, H., Straccia, U.: Information Retrieval and Machine Learning for Probabilistic Schema Matching. Information Processing & Management 43(3), 552–576 (2007)
12. Qian, L., Cafarella, M.J., Jagadish, H.V.: Sample-driven schema mapping. In: Proceedings of the 2012 International Conference on Management of Data, SIGMOD 2012, p. 73. ACM Press (2012)
13. Ravikumar, P., Cohen, W.W.: A Hierarchical Graphical Model for Record Linkage. In: Proceedings of the 20th Conference on Uncertainty in Artificial Intelligence, pp. 454–461. AUAI Press (2004)

14. Leitão, L., Calado, P., Herschel, M.: Efficient and Effective Duplicate Detection in Hierarchical Data. IEEE Transactions on Knowledge and Data Engineering PP(99), 1 (2012)
15. Irani, D., Webb, S., Li, K., Pu, C.: Large Online Social Footprints–An Emerging Threat. In: Proceedings of the International Conference on Computational Science and Engineering. IEEE (2009)
16. Liu, J., Zhang, F., Song, X., Song, Y.I., Lin, C.Y., Hon, H.W.: What's in A Name?: An Unsupervised Approach to Link Users Across Communities. In: Proceedings of the Sixth ACM International Conference on Web Search and Data Mining. ACM (2013)
17. Ji, S., Carin, L.: Cost-sensitive Feature Acquisition and Classification. Pattern Recognition 40(5), 1474–1485 (2007)
18. Ling, C.X., Sheng, V.S., Yang, Q.: Test strategies for cost-sensitive decision trees. IEEE Trans. on Knowl. and Data Eng. 18(8), 1055–1067 (2006)
19. Saar-Tsechansky, M., Melville, P., Provost, F.: Active feature-value acquisition. Manage. Sci. 55(4), 664–684 (2009)
20. Lin, Y.C., Yang, D.N., Chen, M.S.: Selective Data Acquisition for Probabilistic K-nn Query. In: Proceedings of the 19th ACM International Conference on Information and Knowledge Management, pp. 1357–1360. ACM (2010)
21. Tan, Y.F., Kan, M.Y.: Hierarchical cost-sensitive web resource acquisition for record matching. In: Proceedings of the 2010 IEEE/WIC/ACM International Conference on Web Intelligence and Intelligent Agent Technology, vol. 01, pp. 382–389. IEEE Computer Society (2010)
22. Epanechnikov, V.: Non-Parametric Estimation of a Multivariate Probability Density. Theory of Probability & Its Applications 14(1), 153–158 (1969)
23. John, G.H., Langley, P.: Estimating Continuous Distributions in Bayesian Classifiers. In: Proceedings of the Eleventh Conference on Uncertainty in Artificial Intelligence (1995)
24. Zhang, J., Marszalek, M., Lazebnik, S., Schmid, C.: Local Features and Kernels for Classification of Texture and Object Categories: A Comprehensive Study. Int. J. Comput. Vision 73(2) (2007)

Information Diffusion and Influence Measurement Based on Interaction in Microblogging

Miao Yu, Wu Yang, Wei Wang, Guowei Shen, and Guozhong Dong

Information Security Research Center, Harbin Engineering University, Harbin,
Heilongjiang, China 150001
{yumiao,yangwu,w_wei,shenguowei,dongguozhong}@hrbeu.edu.cn

Abstract. In microblogging, user interaction is the main factor that promotes the information diffusion rapidly. According to the user interaction in the process of information diffusion, this paper proposes a directed tree model based on user interaction that considering the history, type and frequency of interaction. User interaction matrix was used to describe the interactions between pairs of users. A directed diffusion tree was generated from the sparsification of interaction graph. The edges of directed diffusion tree were used to measure the information influence and identify the spam in microblogging. Experimental results show that the directed tree model can describe the information diffusion, measure the influence more accurately and identify the spam in the dataset more effectively.

Keywords: Microblogging, Information diffusion model, Influence, User interaction, Measurement matrix.

1 Introduction

As a media platform for rapid information diffusion, Microblogging has become the essential part of life. With the growing popularity of Twitter, Microblogging has been well developed shortly. Till August, 2011, more than 200 million users have registered in Sina Weibo[1]. User can post large number of messages in microblogging, which can be diffused by user interaction. Owing to the fact that the microblogging information spreads fast, but For messages' diffusion is quick and their life cycle is short, it is an important research problem that how to model the message diffusion and measure their influence.

Influence measurement is an important means to analyze microblogging. Researchers are focused on measuring the influence, and finding the influential users [2-4]. The researches considers the user influence mainly through their follow and retweet times. Regarding the retweeting messages, it is hard to measure their influence simply by the displayed data. Changhyun Lee [2] defines the effective readers by message receiving sequence, which can't judge exactly whether user read these messages. Hence, it is a challenge to measure the influence of the user interacts. This paper will consider the historical attributes and life circles of users to measure the influence generated by retweeting the messages.

H.-Y. Huang et al. (Eds.): SMP 2014, CCIS 489, pp. 129–140, 2014.
© Springer-Verlag Berlin Heidelberg 2014

This paper analyses the user interaction behaviors towards one single message on Sina Weibo, to construct a user interaction matrix to measure the variously frequent interactions. Users generate dynamic interaction graph during single message diffusion process, this paper proposed a directed diffusion tree model to describe the information diffusion. The directed diffusion tree is used to find the backbone influential path during the information diffusion process; it enables calculating the influential at any time and identifying the spam by analyzing the directed diffusion tree.

An empirical study of the directed diffusion tree model has been conducted, the results indicate that this tree model can not only dynamically describe the diffusion process, but also analyzes the backbone influential path during the information diffusion process. The diffusion tree model efficiently identifies the spam messages by simply analyzing the directions. Compared to the measurement based on information attributes, the new introduced directed information diffusion tree model is more accurate and convincing.

The rest of this paper is organized as follows. We outline the details of information diffusion, user interaction and influence measurement in section 2. In section 3, we focus on analysis of the user interactions of Sina Weibo, and form a formal description for user interaction. And then we present a dynamic information diffusion model based on user interaction, analysis weight matrix calculation of the model and the processing strategy of sparse of user interaction graph. Empirical analysis of the information diffusion model based on Sina Weibo data set in Section 5. Finally, in Section 6, we conclude and point out some directions for future work.

2   Related Work

Microblogging provides an unprecedented opportunity to the empirical research of social network. Based on the statistical analysis of Twitter network, analyzing and researching the topology characteristics, information diffusion, the finding of influential users and the topic classification[5, 6], the conclusion of follow relation corresponds homogeneity characteristic is drawn.   However, these researches are only in macro topic viewpoints to analysis the large data sets. In 2011, the study on Sina Weibo [7] was firstly made by Louis Yu from the HP Social Computing Lab. Through the contrast analysis of Sina Weibo and Twitter in topics, there is great difference in the information content was found. But authors did not analyze the propagating process of information diffusion and the users' influences to the information diffusion. In this section, three categories of related work were reviewed, such as information diffusion in social network, user behaviors analysis and influence measurement.

Literature [8] proposed a generative model for rapid information diffusion, which includes two independent phase. But it can't describe the information diffusion continuously. Literature [9] researched blog and news on the track of the diffusion path and the effect in the network. In order to simplify the research, author set the network static and unchanged, while the information in mircroblogging is transient. Literature [10] implemented micro and macro research on mail relay networks, considering the user's periodicity. But the user interaction of the microblogging is much more complex and faster than that based on mail. Jure Leskovec [11, 12] carried out the modeling research on information diffusion and proposed a K-tree model to predict message diffusion. The model required the establishment of the actual diffusion process is a balanced tree. In microblogging, information diffusion process is dynamic. Message may explosively spread if the user has a large number of fans. It is very difficult to describe the information diffusion in mircroblogging.

User interaction plays an important role in the process of information diffusion. Researchers have studied on the behaviors of retweet and mention [13, 14, 17]. Michael J.Welch investigated the semantics of the Follow and retweet relationships [15], which can be transferred by retransmission, analyzed the Twitter graph that built by Follow and retweet. But they did not consider the interaction and the time factor of the users.

Influence Measurement. Much work has been focused on influence measurement [16] by comparing the number of comments, retweets, mentions, and reader to measure the influence. In fact, the process of information diffusion is affected by many factors. Single indicator is difficult to measure the influence accurately. Due to the influence during the information diffusion grows and decay rapidly, the influence of the information diffusion process needs to be calculated dynamically.

3 User Interaction Analysis

According to the Chinese Internet development statistical report[1] publish at 2011 July shows that the most popular Chinese microblogging is Sina Weibo, its registered users is more than 200 million till to 2011 August, it provides a researched platform for the analysis of user behavior and the contents of the message, and then we are focus on analysis of user interactions on the Sina Weibo.

3.1 The User Interfaces of Sina Weibo

Although Sina Weibo provides many user operation interfaces, the operation objects only include two entities, including user and message. Operation on user or message entities can generate many types of interaction. Interactive graph as shown in Fig. 1. Yellow solid lines from user to message denote user posting message. Black dotted line from A to B denote user A follow user B. Blue solid line from A to B denote user A comment for user B's message. Red solid line from A to B denote user A's message mention user B. Green solid line from A to B denote user A retweet user B's message. Only interactions between different users are considered here, so include user interaction generated by follow, retweet, mention, comment operations.

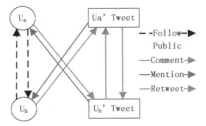

Fig. 1. User interaction diagram

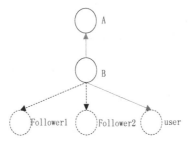

Fig. 2. Interaction diagram representation of the latent and explicit interactions

A user interaction diagram is shown in Fig. 2. The interactions between B and other users are built for B mentioned other users when retweet message. The message M of A is retweeted by B and is sent to F1 and F2 automatically, who are B' fans. Because message sending do not and "F1, F2" can't be determined by obvious content in Sina Weibo need the interaction between F1 and F2, the interaction between B at the time of retweeting. The interactions built by user's retweet affect on his fans are called latent interaction. Those interactions which can be determined by obvious information are called explicit interaction, whose presentation forms include retweet, mention and comment.

If F1 and F2 don't operate the messages retweeted to them, the message influence on them is hard to get. If F1 is the fans of a large number of stars, the message retweeted by B will be overwhelmed by the messages from stars. So latent interactions are determined by the historical interaction attributes between A and "F1, F2"

3.2   Formal Description of User Interactions

The interaction between any two of users in Sina Weibo can be summarized as table 1. User interaction set I = {Follow, Comment, Mention, Retweet}. The Follow and Mention interactions are generated by B act on A, which are passive interaction for A. The Comment and Retweet interaction by B act on A. From causal and temporal, Comment and Retweet interaction will be generated only when A' message already exists.

Table 1. User interactions

	User A	User A' Tweet
User B	Follow	Comment
User B' Tweet	Mention	Retweet

To describe user interaction, interaction measurement matrix is imported, which can show the types of interaction and frequency between users. For A and B, at time t, the interaction measurement matrix for interaction initiated by A is $I^t_{(A,B)} = \begin{bmatrix} Follow_A & Comment_A \\ Mention_A & Retweet_A \end{bmatrix}$ in $(t-1)$ is noted by

$$\Delta I^{(t-1,t)}_{(A,B)} = \begin{bmatrix} \Delta Follow_A & \Delta Comment_A \\ \Delta Mention_A & \Delta Retweet_A \end{bmatrix}$$ and each element is calculated according

to formula 1-4.

$$\Delta Follow_A = \omega_{Follow} \tag{1}$$

$$\Delta Comment_A = \sum_{(t-1,t)} \omega_{comment} \tag{2}$$

$$\Delta Mention_A = \sum_{(t-1,t)} \omega_{Mention} \tag{3}$$

$$\Delta Retweet_A = \sum_{(t-1,t)} \omega_{Retweet} \tag{4}$$

For the different types of interaction, the weight relationship is $\omega_{Retweet} \geq \omega_{Comment} \geq \omega_{Mention} \geq \omega_{Follow}$

4   Information Diffusion Model

In Sina Weibo, The diffusion processes are different for different message under the various interactions. The computation and decay of weight of user interaction are introduced firstly, then the information diffusion directed tree model building process is described, based on sparsification of user interaction graph.

4.1   Computation of Interaction Weight

According to user interaction measurement matrix, for a interaction initiated by one user, its weight is related with not only the types and frequency of interaction, but also the user's overall attribute. The interactions generated by different users have different effects on information diffusion, which is described by historical attributes coefficient of user corresponding to different interaction. For a user A, his historical follow coefficient is ω^A_{Follow}, his historical comment coefficient is $\omega^A_{Comment}$, his historical mention coefficient is $\omega^A_{Mention}$, and his historical retweet coefficient is $\omega^A_{Retweet}$.

So after introduction of user historical attribute weight, the weight of some interactions initiated by A in $(t\text{-}1,t)$ can be calculated by formula 5-8.

$$\Delta Follow_A = \omega_{Follow} \times \omega_{Follow}^A \tag{5}$$

$$\Delta Comment_A = \sum_{(t-1,t)} \omega_{Comment} \times \omega_{Comment}^A \tag{6}$$

$$\Delta Mention_A = \sum_{(t-1,t)} \omega_{Mention} \times \omega_{Mention}^A \tag{7}$$

$$\Delta Retweet_A = \sum_{(t-1,t)} \omega_{Retweet} \times \omega_{Retweet}^A \tag{8}$$

From statistical view, user behavior present periodicity, the effect of different interaction is different in different period, which is described by user influence function. The period of this function is day. F_A is the effect function of user A, F_B is the effect function of user B. For the Follow and Comment interaction initiated by A, they are acting on B' message entities. So these interactions have relationship with influence of user B. For the Mention and Retweet interaction initiated by A, they are acting on A' message entities, so these interactions have relationship with influence of user A. After the introduction of user influence function, the weight of user A' interaction is $I_{(A,B,T)}^t$, in $(t\text{-}1,t)$, the weight of interaction initiated by A is $\Delta I_{(A,B,T)}^{(t\text{-}1,t)}$.

$$\Delta I_{(A,B,T)}^{(t\text{-}1,t)} = \begin{bmatrix} F_A \\ F_B \end{bmatrix}^T \cdot \Delta I_{(A,B)}^{(t\text{-}1,t)} \tag{9}$$

4.2 Computation of Interaction Weight Decay

The life cycle of message is normally short in Sina Weibo. In the information diffusion process, the message influence will decay quickly with time goes by, the same way to interaction weight. Because the weight of interaction acting on users decay very slowly, but decay quickly for that acting on messages, the decay of different interaction will be described respectively. For the interaction weight of user A at time t-1 is $I_{(A,B,T)}^{t\text{-}1} = \begin{bmatrix} I_{user} & I_{Teweet} \end{bmatrix}$, I_{user} is the influence of interaction acting on user entity, and is the influence of interaction acting on message entity. So the decay matrix is $D = \begin{bmatrix} D_{user} \\ D_{Teweet} \end{bmatrix}$, D_{user} is the decay coefficient for users, and D_{Teweet} is the decay coefficient for message. At time t, user interaction matrix is recorded as $\omega_{(A,B)}^t$, and the weight of user interaction is calculated by formula 10, $I_{(A,B,T)}^{t\text{-}1} \cdot D$ is the user interaction weight at time t-1. As user interactions are newly in time $(t\text{-}1,t)$, so user interaction weight is $\Delta I_{(A,B,T)}^{(t\text{-}1,t)} \cdot \begin{bmatrix} 1 \\ 1 \end{bmatrix}$.

$$\omega_{(A,B)}^t = I_{(A,B,T)}^{t\text{-}1} \cdot D + \Delta I_{(A,B,T)}^{(t\text{-}1,t)} \cdot \begin{bmatrix} 1 \\ 1 \end{bmatrix} \tag{10}$$

4.3   Building Directed Diffusion Tree

At time t, information diffusion process is directed diffusion tree $C^t = (V, E, \omega)$, here V is the set of user attending interaction in single message diffusion process. E is the directed edge representing user interaction. ω is the weight of interaction between two users. The building process of directed diffusion tree includes three steps:

- Build the user interaction graph for single message diffusion at time t.
- Calculate weight of user interaction according to user interaction behaviors.
- A weighted directed diffusion tree is built according to the sparsification of edges in user interaction graph. Here the user interaction graph is a complex cycled graph, which can't describe information diffusion accurately. Two sparsification processes are shown in Fig.3. In Fig. 3(a), the left part is the original user interaction graph, at time t_2 , user B retweets user A's message and mention user C, which forms a triangle cycle. In Fig. 3(b), at time t_4 , user B retweets the message back from D, which makes a iteration.

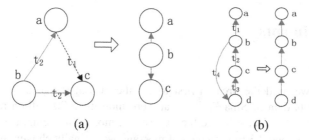

Fig. 3. sparsification of user interactions

A dynamic sparsification strategy is used to build information diffusion tree based on user interaction graph. This strategy can guarantee that the diffusion tree structure can be always held and only one edge is used to describe information diffusion between users. For the interaction initiated by A acting on B, B is in tree at time t-1, and a directed edge exists between B and C. According the strategy, an interaction edge between A and C is generated at t_1 in graph A in Fig.3, so compare the weight directly. Because the Mention weight is heavier that Follow weight, the BC is held, AC is deleted. The sparsification result is shown in the right part in Fig.3(a). For Fig.3(b), the directed edge exists between D and C at time t_3 . Comparing the depth of B and C at t_3 , for the depth of B is smaller than C, so the edge added newly is deleted. The sparsification result is shown in the right part in Fig. 3(b).

User N_1 publish a message M, the corresponding diffusion tree is shown in Fig.4. At any time, the information diffusion process corresponds to a directed diffusion tree, this model can describe the join and exit of users based on single message, depict all the interaction processes between users base on single message, and analysis the backbone of information diffusion path.

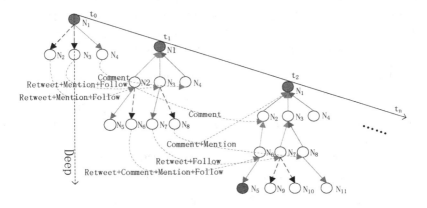

Fig. 4. Information diffusion tree model

5   Experiments

5.1   Dataset

In this paper, we used the breadth-first to get the messages that users released and related comments data of Sina Weibo that were more than 70 million messages and 177 million users. In order to make the experiment more convincing, the original data sets were screened and 2.3 million data of message were finally chosen, which include about 100 million comments. To protect the user's privacy, the privacy protection was used to deal with the data set. In order to hide the original user ID information, the data set used the hash to deal with the real user ID. Experimental data set did not involve any user-released messages and user-related personal privacy information.

By using user interaction, microblogging spread the message fast. Retransmission and comments are two important interactions of the microblogging users, which were used to statistics for the time differences of retransmission and comments. Fig5 make a statistics for delay time of the message retransmission and comments to review interactive statistical distribution in the life cycle of 10 days. Fig 5 (a) shows that the delay time of retweeting the message is very short. Fig 5(b) shows that the delay time of the comments that users can see is very short, and even shorter than that of retweet, which is achieved with Sina Weibo's mechanism. Pop-up window was used to prompt the user about the comments, which allowed the user to have a timely response.

Due to the impact of the real life, the quick join and exit a single message diffusion process of the Microblogging users show cyclical fluctuation. In figure 6, message-based user interactions comply with the period of the day, and the user interactions are concentrated in the morning and evening. Information retransmission is the main interactions of the information diffusion process, however, the comments of the information diffusion process also played an important role in information diffusion.

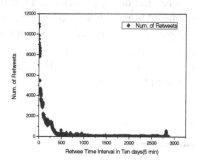
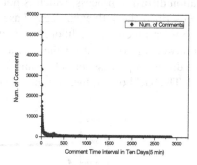

(a) The distribution of user retweet   (b) The distribution of user comment

Fig. 5. The distribution of user interaction decay

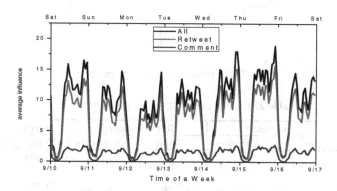

Fig. 6. The cycle distribution of user interaction

5.2   Influence Measurement

In the whole message diffusion process, the influence of the message is a rapid changed, which is affected by various factors. The paper used the interaction matrix of users to measure the influence of the message. By traversing the side of the directed tree, the influence of the message at any time can be got, while the accumulation of the weight of the side is the message influence. The number of follow and message diffusion are commonly used to measure the influence of the message, however, the influence of the message in the message transmission process also influenced by other factors. Table 2 shows three method of influence measurements upon the data sets of Sina Weibo, and the influence measurements based on user interaction did not consider the decay process.

In table 2, we can find that the measurement based on user interaction is more accurately, the diffusion paths of the topic of "Qinghai TV" have large number of followers, but they don't retweet or comment the message. The diffusion process of the single message as show in Fig 7, the black line is described the influence in information diffusion process, which is not considered the decay process, and the red line is described the influence decay based on time. After 24 hours, the message's influence wasn't increased, Although the diffusion path have some star users, but they did not retweet or comment for the message. So our model can not only describe the message diffusion process, but also can find the reason of the message influence was changed. The backbone influence users are described in Section 5.3.

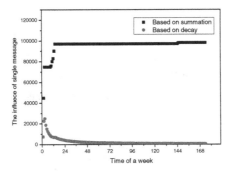

Fig. 7. The influence of a single message

Table 2. The compare of the influence measurement

Rank	Follow-based		Retweet-based		Interaction-based	
	Topic	**Follower**	**Topic**	**Retweets**	**Topic**	**Influence**
1	Qinghai TV	55175530	Weight loss1	52316	Weight loss1	60154
2	Compensation	36583147	Magic weight	31589	Magic weight	28404
3	Government	31310643	Star	14956	Qinghai TV	47164
4	Hugh Jackman	30224400	Weight loss2	14409	Weight loss2	12524
5	Children	28936112	Taobao	14185	dress	12414
6	Constellation	27362742	Constellation	13888	star	11994
7	married	25478217	Medical	13090	Medical	9870
8	consumption	23032075	Obesity	13019	Joke	9174
9	Customer	20358897	dress	12879	Taobao	9084
10	Lan Sang	20088000	Joke	12248	Medical	9084

5.3   Analysis of the Diffusion Process

Microblogging network released a large number of messages everyday, which contains some auto-reply spam messages. It is difficult to identify spam messages based on the content of the message, however, the behavior of garbage generated messages can be effectively identified. We can analyze the spam messages by analyzing the directed tree of information diffusion. Table2 shows the identified experimental data set of spam messages. Spam messages mainly contains the merchandising of Taobao, website promotion, voting, winning, etc. The content of spam messages have no semantic relevance with the content of the message itself.

Table 3 lists three backbone influence users in information diffusion. From the main interaction attributes of these three users, including follow, Comment and Retweet, the value of these attributes are consistent with their influence state. Table 3 testifies the accuracy of the backbone influence user discovery strategy.

Table 3. Backbone influence users

User Name	Followers	Comments	Retweets
Chen Yao	14271359	1659	6505
Chen Li	1664999	214	1179
AnnXiao	23818	1630	256

Table 4 shows the spam identification result of the data set. The first column of Table 4 lists the topics of original messages. The second column list the topics of spam identified from the comment of the original messages. As the topics shown, the topics of spam have nothing to do with those of original messages, which testifies the effectiveness of the propose spam detection based on information diffusion model.

Table 4. Spam category in Sina Weibo data set

Original Message(topic)	Spam(topic)
Brain teasers	Taobao, Promotion
Lose weight	Requested to follow
Gourmet food	Website promotion
Birthday	Vote
News	Gifts

6   Conclusion

The paper focuses on the analysis of user interactions, measures the variously frequent interactions by using the interaction metrics. The dynamic information diffusion tree model has been proposed based on the interaction matrix. The diffusion tree model was used to analyze the information diffusion process, measures its dynamic influence.

Our experiments show that the way to measure the information influence by considering the user interactions is more accurate than simply considering the number of followers or retweets. The directed tree model was used to describe the information diffusion process, and the backbone influence path was found in the single information diffusion. According to the user interactions of the diffusion tree model, we identified some spam

Acknowledgments. The work is supported by the Natioal Natural Science Foundation of China(61170242), 863 High Technology Program(2012AA012802), and Fundamental Research Funds for the Central Universities(HEUCF100611).

References

1. CNNIC. The 28th statistics report on the internet development in China (2011)
2. Lee, C., Kwak, H., Park, H., Moon, S.: Finding Influentials Based on the Temporal Order of Information Adoption in Twitter. In: WWW 2010, pp. 1137–1138 (2010)
3. Cha, M., Haddadi, H., Benevenuto, F., Gummadi, K.P.: Measuring User Influence in Twitter: The Million Follower Fallacy. In: WOSA 2010 (2010)
4. Romero, D.M., Galuba, W., Asur, S., Huberman, B.A.: Influence and Passivity in Social Media. In: WWW 2011, pp. 113–114 (2011)
5. Kwak, H., Lee, C., Park, H., Moon, S.: What is Twitter, a Social Network or a News Media? In: WWW 2010, pp. 591–600 (2010)
6. Wu, S., Hofman, J.M., Mason, W.A., Watts, D.J.: Who Says What to Whom on Twitter. In: WWW 2011, pp. 705–714 (2011)
7. Yu, L., Asur, S., Huberman, B.A.: What Trends in Chinese Social Media. In: SNA-KDD 2011 (2011)
8. Dyagilev, K., Mannor, S., Yom-Tov, E.: Generative Models for Rapid Information Propagation. In: SOMA 2010, pp. 35–43 (2010)
9. Gomez-Rodriguez, M., Leskovec, J., Krause, A.: Inferring Networks of Diffusion and Influence. In: KDD 2011, pp. 1019–1028 (2011)
10. Wang, D., Wen, Z., Tong, H., Lin, C.-Y., Song, C., Barabási, A.-L.: Information Spreading in Context. In: WWW 2011, pp. 735–744 (2011)
11. Sadikov, E., Medina, M., Leskovec, J., Garcia-Molina, H.: Correcting for Missing Data in Information Cascades. In: WSDM 2011, pp. 55–64 (2011)
12. Yang, J., Leskovec, J.: Modeling Information Diffusion in Implicit Networks. In: IEEE International Conference on Data Mining, pp. 599–608 (2010)
13. Ghosh, R., Surachawala, T., Lerman, K.: Entropy-based Classification of 'Retweeting' Activity on Twitter. In: KDD 2010(2010)
14. Hong, L., Dan, O., Davison, B.D.: Predicting Popular Messages in Twitter. In: WWW 2011, pp. 57–58 (2011)
15. Welch, M.J., Schonfeld, U., He, D., Cho, J.: Topical Semantics of Twitter Links. In: WSDM, pp. 327–336 (2011)
16. Lee, C., Kwak, H., Park, H., Moon, S.: Finding Influentials Based on the Temporal Order of Information Adoption in Twitter. In: WWW 2010, pp. 1137–1138 (2010)
17. Itakura, K.Y., Sonehara, N.: Using Twitter's Mentions for Efficient Emergency Message Propagation. In: ARES 2013, pp. 530–537 (2013)

Inferring Emotions from Social Images Leveraging Influence Analysis

Boya Wu[1,2], Jia Jia[1,2], Xiaohui Wang[1,2], Yang Yang[1], and Lianhong Cai[1,2]

[1] Department of Computer Science and Technology, Tsinghua University, Beijing 100084, China
[2] Key Laboratory of Pervasive Computing, Ministry of Education,
Tsinghua National Laboratory for Information Science and Technology (TNList)
stella.1991@163.com, jjia@mail.tsinghua.edu.cn,
wangxh09@mails.tsinghua.edu.cn, sherlockbourne@gmail.com,
clh-dcs@tsinghua.edu.cn

Abstract. Nowadays thriving image-based social networks such as Flickr and Instagram are attracting more and more people's attention. When it comes to inferring emotions from images, previous researches mainly focus on the extraction of effective image features. However, in the context of social networks, the user's emotional state is no longer isolated, but influenced by her friends. In this paper, we aim to infer emotions from social images leveraging influence analysis. We first explore several interesting psychological phenomena on the world's largest image-sharing website Flickr[1]. Then we summarize these pattern into formalized factor functions. Introducing these factors into modeling, we propose a partially-labeled factor graph model to infer the emotions of social images. The experimental results shows a 23.71% promotion compared with Naïve Bayesian method and a 21.83% promotion compared with Support Vector Machine (SVM) method under the evaluation of F1-Measure, which validates the effectiveness of our method.

Keywords: Images, emotion, social influence.

1 Introduction

Emotion plays a vital role in human life. It is said that emotion stimulates the mind 3,000 times faster than rational thoughts[1]. Understanding the internal dynamics and external manifestation of human emotion can not only pushes forward the frontier of sociology and psychology, but also benefits the promotion of product design and user experience.

The factors that determine human emotional state varied. We may be delighted for a feast or a beautiful melody, and we may feel sad because of our friends' bitter experience. The way we express our feelings also varies. Besides texts, images provide us with a novel way to convey what we feel. Compared with texts, images are more vivid, freer, but more difficult and more subjective to understand as well.

---

[1] www.flickr.com

H.-Y. Huang et al. (Eds.): SMP 2014, CCIS 489, pp. 141–154, 2014.
© Springer-Verlag Berlin Heidelberg 2014

Nowadays, with the thriving of many image-based social networks, such as Flickr and Instagram, we're used to upload images on the Internet and share them with our friends. The rapid development of these websites and applications provides us with large amounts of available data, thus offering us a great opportunity to study the problem of the inference of emotions embedded in images.

Previous researches mainly focus on the exploration of effective image features, such as colors, composition and shapes. J.Machajdik and A.Hanbury[2] pick up four categories of low-level features like wavelet textures and GLCM-features and proved the effectiveness of these features on three data sets. Moreover, from an aesthetic point of view, X.Wang[3] explored the interpretable image features, including the color combinations, saturation, brightness, the warm and cool color ratio, etc. and conduct experiments on two public data sets. Other works can be found in [14], [15]. However, in the context of social networks, the user's emotion state is no longer isolated. That is to say, the user's emotional state is not only determined by images features, but also affected by her friends' emotional states and her unique experience.

This point of view is also supported by sociological and psychological theories. S.Hareli and A.Rafaeli[4] discovered that one's emotion will influence other people's emotion, thoughts and behavior. In turn, other people's reaction will influence the future interaction between them as well as this person's future emotion and behavior. J.Fowler and N.Christakis[5] studied social networks and mentioned that people's happiness depends on the happiness of those who are connected to them.

In this paper, we aim to tackle the image emotion inference problem by leveraging the influence analysis. Figure 1 illustrates the procedure of our work. Our study is based on the world's largest image-sharing website Flickr and 2,060,353 images and 1,255,478 user profiles are constructed as the data set of our experiment. Based on these data, three types of correlations are considered and studied, namely, 1) Attributes correlation: the correlation between image emotion and image features. 2) Temporal correlation: the correlation between the emotion of current image and the images the user uploaded before. 3) Social correlation: the correlation between the image emotion and the user's interaction with her friends. We first make observations and discover several interesting psychological phenomena about the emotion influence on the social network and then summarize them into formalized factor functions. Next by introducing these factor functions we propose a partially-labeled factor graph model which fulfils the inference of images emotion. The experimental results show that the F1-Measure reaches 0.4251 on average, which increases by 23.71% compared to Naïve Bayes and 21.83% compared to SVM, thus validating the effectiveness of our method.

The rest of this paper is organized as follows: Section 2 presents the data observation. Section 3 formulates the problem. Section 4 explains the proposed model and algorithm. Section 5 shows the experiments we conduct and investigates the experimental results. Section 6 concludes the work of this paper.

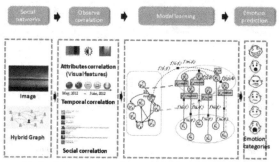

Fig. 1. The procedure of inferring emotions from social images leveraging influence analysis

2   Data Observation

In order to leverage the information of emotion influence, to begin with we need to explore the patterns emotion influence acts on the social networks. In this work, we direct our attention to the following three aspects: the intimacy between the user and her friends, the dominant emotion of friends and the composition of friends.

2.1   Data

We randomly downloaded 2,060,353 images and 1,255,478 users from Flickr and employ them as our data set. To evaluate the performance of inferring emotions, we first need to know the primitive emotions embedded in the images. Due to the large-scale of our data set, manually labeling the emotion for every image is not practical. Therefore we adopt a method used by Y.Yang[12] to automatically label the emotions. Based on Ekman's theory[6] that human emotion can be generally classified into six categories, namely, happiness, surprise, anger, disgust, fear and sadness, we first construct six word list in which words are closely related to the basic emotions from WordNet[7] and HowNet[8]. Then we manually verify every word list to make sure that the words are relative to the emotions. Next we compare every word in the tag written by the user who upload the image and find out which word list matches the tag the best. The image is labeled with that basic emotion if the best match exists. By this method, 354,192 images are qualified. To make sure that the data we used for observation is typical and representative, we further restrict the completeness of data, which means that the time stamp of the image and the related information is not absent. In this way, we pick out 10,448 images for data observation.

2.2   The Intimacy between Friends

The intimacy between friends may exert a profound influence on the degree of emotion influence. For example, the user may experience the same emotion as her best friend, but may not feel linked with a newly acquainted person. Because comments of

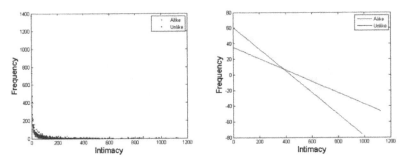

Fig. 2. The correlation between emotion influence and the intimacy between friends

images are public to every user, so herein we define friends as the other users whose images are ever commented by this user. Meanwhile, we carefully make an assumption that the user's emotional state can be conveyed through the image the user uploads. Given the definition and the assumption, we observe the emotional states of users and their friends on the day they upload an image and the day before. We also observe the interactions between users and their friends.

Figure 2 illustrates the observation results. It is noted that when the user is intimate with a friend, she probably feel the same with that friend, but when the user is not familiar with a friend, she may not likely to influenced by that friend.

2.3   The Dominant Emotion of Friends

The dominant emotion of the user's friends may affect the user's emotional state. Interestingly, will the user feel happy if all her friends feel happy? Will the answer remain the same when it comes to different emotion categories? To find answers of these questions, we conduct experiments and the observation results are shown in Figure 3.

As can be seen from Figure 3, whether we consider emotion influence or not, over half of the users on the image-based social networks are happy. What's more, negative emotions, namely, surprise, anger, disgust, fear and sadness are more capable of influencing others. In particular, when most of the user's friends feel surprised, angry or disgust, the possibility of the user feel the same doubled or even trebled. However, in terms of positive emotion, which refers to happy in the six basic emotion categories, the tendency is not obvious, indicating that negative emotions have a rather larger influence than positive emotions.

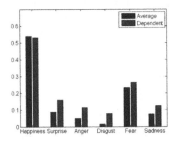

Fig. 3. The correlation between emotion influence and the dominant emotion of friends

2.4   The Composition of Friends

The composition of friends may have something to do with the degree of emotion influence. For instance, are those having friends of various locations or occupations more likely to be influenced or those having similar friends? Are those having more friend inclined to be influenced or those having less friends?

Excitingly, several interesting as well as important phenomena are discovered. The left figure in Figure 4 shows that users who have less friends are more likely to be influenced by their friends and users who have more friends are less likely to be influenced, while no obvious difference is found in terms of the location diversity and occupation diversity. The middle figure indicates that when users have more female friends, they are more probable to be affected. The right figure is a little bit different. It's not about the users' friends, but the users themselves. Interestingly, it can be seen that female users and taken users have a tendency to feel the same with their friends compared with male users and single users.

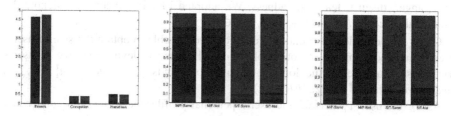

Fig. 4. The correlation between emotion influence and the composition of friends

3   Problem Definition

By exploring the patterns of emotion influence, we discovered many interesting psychological phenomena. In this section, we'll define the notations of this work and formalize the problem we study.

The goal of our work is to analyze and predict the emotion of images uploaded by users in the context of social networks leveraging influence analysis.

A static social network can be represented as $G = (V, E)$ where V is the set of $|V| = n$ users, $E \subset V \times V$ is the set of relationship among users. In our work, this relationship is can be regarded as the friendship. First, we should give friend a definition.

Definition 1. Friend: If user v_i ever comments any image uploaded by user v_j before time t, then v_j is called v_i's friend and v_i's friends at time t are denoted as $FR^t(v_i)$.

Based on the definition, it can be seen that user v_i's friends $FR^t(v_i)$ at time t may be empty, which means that v_i has never commented any image before. As time goes by, when v_i makes her first comment, $FR^t(v_i)$ won't be empty anymore. So $FR^t(v_i)$ is time-dependent, and more precisely E is better denoted as E^t.

Acquiring the concept of friend, we can further define the intimacy between friends.

Definition 2. Intimacy: After standardization, the frequency of user v_i making comments of v_j's images at time t denoted as μ_{ij}^t is called the intimacy between user v_i and v_j.

The standardization here is the process of discretizing the frequency into three degrees of intimacy, namely, $\mu_{ij}^t \in \{1,2,3\}$. The larger μ_{ij}^t is, the higher intimacy between user v_i and v_j is.

On the base of these definitions, we define emotion next.

Definition 3. Emotion: The emotional state of user v_i at time t is denoted as y_i^t. The emotion of the image x_{ij}^t uploaded by user v_i at time t is denoted as y_{ij}^t. In our work, an assumption is carefully made as follows: According to Ekman's theory, human emotion can be classified into six basic categories, which are happiness, surprise, anger, disgust, fear and sadness. We adopted Ekman's theory and denote the emotional space as R, where $y_i^t \in R$, $y_{ij}^t \in R$.

Given the definition of friend and emotion, we are able to obtain the size of user's friends and the dominant emotion of friends. Herein we devote the size of user's friends as s_i^t and denote the dominant emotion of friends as m_i^t, where $m_i^t \in R$. It is clearly noted that $s_i^t = size(FR^t(v_i))$.

As for the user v_i herself, the user's personal profile, including v_i's gender and whether v_i is single or taken, can be defined as vector p_i. p_i is stable.

Besides, we use X to represent all images uploaded by all users from V. In detail, we use X^L to represent the subset of X where the images are labeled, and use X^U to represent the subset of X where the images are unlabeled. For an image $x_{ij}^t \in X$ uploaded by user v_i at time t, the image features are denoted as vector u_{ij}^t.

We can see that the relationship between users are varying with time. Therefore a time-varying social network is to be defined.

Definition 4 Partially-Labeled Time-Varying Social Network: A partially-labeled time-varying social network can be defined as $G = (V, E^t, X^L, X^U)$, where V is the set of users, E^t represents the friendship between users at time t, X^L represents the labeled images and X^U represents the unlabeled images.

Accomplishing the definition of these notations, the learning task of our model is put forward as follows.

Learning Task: Given a partially-labeled time-varying social network $G = (V, E^t, X^L, X^U)$, find a function f to predict the emotion of all unlabeled images:

$$f: G = (V, E^t, X^L, X^U) \rightarrow Y \tag{1}$$

where $Y = \{y_{ij}^t\}$, $y_{ij}^t \in R$, meaning the emotion of images.

4   Model

In this paper, we proposed a factor graph model to infer the emotion of social images. In the model, many types of correlations are defined and introduced into the model through factor functions. These correlations are: 1) Attributes correlation: The emotion of the image is basically induced by the image features. 2) Temporal correlation: The emotion of current image may be related to the emotion of images the user upload before. 3) Social correlation: The emotion embedded in the image may be influenced by the user's interaction with her friends. These correlations can better help us with the emotion inference problem.

4.1   The Predictive Model

The graphical representation of the model is illustrated in Figure 2. The model is constructed on a basic factor graph model. The input of the model is a partially-labeled time-varying network G, and after the learning and training process, the emotions of the unlabeled images are inferred effectively.

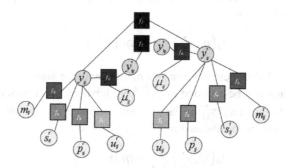

Fig. 5. The illustration of the model

There are two types of nodes in the factor graph model, one called variable node and the other called function node. Shown in Figure 2, the variable nodes are round and the function nodes are square. In our work, all images uploaded by users and their relative information can be regarded as variable nodes and the attributes correlation, temporal correlation and social correlation are leveraged in the form of formalized factor functions.

In the model, the three correlations mentioned above are defined as follows.

1) Attributes correlation: The attributes correlation refers to the correlation between image emotion and image features. It can be encoded as $f_1(u_{ij}^t, y_{ij}^t)$, where u_{ij}^t means the image features of image x_{ij}^t uploaded by user v_i at time t, y_{ij}^t means the emotion of image x_{ij}^t. By introducing parameter α, it can be instantiated as exponential-liner function:

$$f_1(u_{ij}^t, y_{ij}^t) = \frac{1}{z_\alpha} exp\{\alpha^T \cdot u_{ij}^t\} \tag{2}$$

2) Temporal correlation: The temporal correlation indicates the correlation between the emotion of current image and emotions of images the user uploaded before. It can be depicted as $f_2(y_i^{t'}, y_i^t)$, $t' < t$, where y_i^t and $y_i^{t'}$ means the emotion of user v_i at time t and t', conveyed by the images the user uploaded at that time. By introducing the function $g(y_i^{t'}, y_i^t)$ to depict the correlation, it can be instantiated as follows:

$$f_2(y_i^{t'}, y_i^t) = \frac{1}{z_\xi} exp\{\xi_i \cdot g(y_i^{t'}, y_i^t)\} \tag{3}$$

3) Social correlation: The social correlation means the correlation between the emotion embedded in the image and the user's interaction with her friends. Based on the data observation, the social correlation can be detailed into following four aspects.

➤   The correlation between image emotion and the user's personal attributes: The correlation can be encoded as $f_3(p_i, y_{ij}^t)$, where p_i is the personal information of user v_i, including v_i's gender and whether v_i is single or taken, and y_{ij}^t is the emotion of image x_{ij}^t. By introducing parameter β, it can be instantiated as:

$$f_3(p_i, y_{ij}^t) = \frac{1}{z_\beta} exp\{\beta^T \cdot p_i\} \tag{4}$$

➤   The correlation between image emotion and the size of user's friends: The correlation can be encoded as $f_4(s_i^t, y_{ij}^t)$, where s_i^t represents the size of user v_i's friends and y_{ij}^t is the emotion of image x_{ij}^t. By introducing parameter γ, it can be instantiated as:

$$f_4(s_i^t, y_{ij}^t) = \frac{1}{z_\gamma} exp\{\gamma^T \cdot s_i^t\} \tag{5}$$

➤   The correlation between image emotion and the dominant emotion of the user's friends: The correlation can be encoded as $f_5(m_i^t, y_{ij}^t)$, where m_i^t represents the dominant emotion of user v_i's friends and y_{ij}^t is the emotion of image x_{ij}^t. By introducing parameter δ, it can be instantiated as:

$$f_5(m_i^t, y_{ij}^t) = \frac{1}{z_\delta} exp\{\delta^T \cdot m_i^t\} \tag{6}$$

➤   The correlation between image emotion and the user's intimacy with her friends: The correlation can be encoded as $f_6(y_i^t, y_j^t, \mu_{ij}^t)$, where y_i^t and y_j^t represents the emotional states of user v_i and v_j at time t and μ_{ij}^t measures the intimacy between user v_i and v_j at time t. By introducing the function $h(y_i^t, y_j^t, \mu_{ij}^t)$ to depict the correlation, it can be instantiated as follows:

$$f_6(y_i^t, y_j^t, \mu_{ij}^t) = \frac{1}{z_\eta} exp\{\eta_{ij} \cdot h(y_i^t, y_j^t, \mu_{ij}^t)\} \tag{7}$$

So far we have formalized all the correlations in the model. Thus we can define the joint distribution of the model:

$$P(Y|G) = \frac{1}{Z} \times \prod_{x_{ij}^t \in X, v_i \in V} f_1(u_{ij}^t, y_{ij}^t) \times \prod_{x_{ij}^t \in X, v_i \in V} \prod_{y_i^{t'} \in SU^t(v_i)} f_2(y_i^{t'}, y_i^t)$$

$$\times \prod_{x_{ij}^t \in X, v_i \in V} f_3(p_i, y_{ij}^t) \times \prod_{x_{ij}^t \in X, v_i \in V} f_4(s_i^t, y_{ij}^t) \times \prod_{x_{ij}^t \in X, v_i \in V} f_5(m_i^t, y_{ij}^t)$$

$$\times \prod_{x_{ij}^t \in X, v_i \in V} \prod_{v_j \in FR^t(v_i)} f_6(y_i^t, y_j^t, \mu_{ij}^t) = \frac{1}{Z} exp\{\theta^T S\} \qquad (8)$$

where $Z = Z_\alpha Z_\xi Z_\beta Z_\gamma Z_\delta Z_\eta$ is the normalization term, S is the aggregation of factor functions over all nodes, θ denotes all the parameters, i.e., $\theta = \{\alpha, \beta, \gamma, \delta, \xi_i, \eta_{ij}\}$, and $Y = \{y_{ij}^t\}$, $y_{ij}^t \in R$, meaning the inferred emotion for images.

Therefore the target of the inference process is to maximize the log-likelihood objective function $O = log\, P(Y|G)$, and the keynote of the training process is to learn $\theta^* = arg\, max\, O(\theta)$.

4.2   Model Learning

Given the model's input and output, we make clear the parameters and the objective function of the model. Next we'll detail the learning process of the model and the algorithm is summarized as follows.

Algorithm: the learning and inference algorithm of image emotion

Input: a partially-labeled time-varying network $G = (V, E^t, X^L, X^U)$, learning ratio λ

Output: parameter group $\theta = \{\alpha, \beta, \gamma, \delta, \xi_i, \eta_{ij}\}$, inference result $Y = \{y_{ij}^t\}$, $y_{ij}^t \in R$

Read in network G. Construct factor graph. Set up variable nodes and function nodes.
Initiate parameters $\theta = \{\alpha, \beta, \gamma, \delta, \xi_i, \eta_{ij}\}$

Repeat

        Calculate $E_{p_\theta(Y|Y^U, G)}S$ using LBP

        Calculate $E_{p_\theta(Y|G)}S$ using LBP

        Calculate the gradient of θ: $E_{p_\theta(Y|Y^U, G)}S - E_{p_\theta(Y|G)}S$

        Update θ with the learning ratio λ: $\theta = \theta_0 + \frac{\partial O}{\partial \theta} \cdot \lambda$

Until convergence

Get the inference result $Y = \{y_{ij}^t\}$, $y_{ij}^t \in R$ and the trained parameters $\theta = \{\alpha, \beta, \gamma, \delta, \xi_i, \eta_{ij}\}$

To observe which part each parameter is linked to in the objective function, $O = log\, P(Y|G)$ can be rewritten as follows:

$$O = log\, P(Y|G) = log \sum_{Y|Y^U} exp\{\theta^T S\} - logZ$$

$$= log \sum_{Y|Y^U} exp\{\theta^T S\} - log \sum_{Y} exp\{\theta^T S\} \qquad (9)$$

Thus the gradient of the parameter can be obtained. The gradient of θ can be represented as:

$$\frac{\partial O}{\partial \theta} = \frac{\partial(\log \Sigma_{Y|Y^U} exp\{\theta^T S\} - \log \Sigma_Y exp\{\theta^T S\})}{\partial \theta} = E_{p_\theta(Y|Y^U,G)}S - E_{p_\theta(Y|G)}S \qquad (10)$$

Then we can update θ by $\theta = \theta_0 + \frac{\partial O}{\partial \theta} \cdot \lambda$ according to the learning ratio λ. λ can be manually defined and in our work $\lambda = 0.1$.

Given the current parameters, we can obtain the inferred emotion of unlabeled images which fulfills the target of modeling.

5   Experiments

In this section, we validate the effectiveness of the proposed method through experiments. First we give an illustration of the data set we employed. Then we compare the experimental results with other baseline methods. Finally to find out which factor contributes the most, we conduct factor contribution analysis and present some interesting sociological and psychological phenomena.

5.1   Experimental Setup

The data set we employed is randomly downloaded from world's largest image-sharing website Flickr which comprised of 2,060,353 images and 1,255,478 users. As explained in Section 2, by extracting emotion from image tags and checking the completeness of relative information, 218,816 qualified images are left. In order to test the performance of every emotion category, we evenly and randomly pick out 6,000 images from every emotion category and 36,000 images are chosen in total.

As for image features, considering the interpretability of image features and the fine performance when used for emotion inference, we adopt the method proposed by X.Wang[3] to extract the features, including the five major color types, saturation, brightness and so on.

We conduct performance comparison experiments to demonstrate the effectiveness of our method. The two baseline methods we employed for comparison are the Naïve Bayesian method and the SVM method.

Naïve Bayesian: Naïve Bayesian is a widely used classifier. The image features are directly inputted into the classifier and the classifier outputs the inferred image emotions.

SVM: SVM is frequently used in many classification problems. Here it directly uses the image features as the input and outputs the inferred image emotion. In this work, we use LIBSVM design by C.Chang and C.Lin[9].

The data set we employed to the baseline methods and our proposed methods is the same, 60% for training and 40% for testing as well. The evaluation measure we used contains precision, recall and F1-Measure.

5.2   Experimental Results

Table 1 exhibits the experimental results.

From the table, it is apparent that our method significantly enhanced the performance of the image-based emotion inference problem. The average F1-Measure reaches 0.4254, increased by 23.71% compared with Naïve Bayesian and 21.83% compared with SVM. The Naïve Bayesian method and the SVM method barely use the image features for training and let go of the emotion influence on the social network. Under this circumstance, the classifier can only try to analyze the image emotion from the image's saturation, brightness and so on. Though these methods are effective to a certain extent, in the complex context of social networks, the user's emotional state is no longer isolated and influenced by other users' emotional states instead. So taking only the image features into account cannot measure the user's emotional state precisely. In contrast, because our method concerns about the emotion influence on the social network and formalize these patterns into the factor graph model, it better depicts the real situation of the social networks and examines the user's emotional state in its entirety, thus remarkably promote the performance of the emotion inference.

Table 1. Performance of emotion inference

Emotion	Method	Precision	Recall	F1-Measure
Happiness	Naïve Bayesian	0.2026	0.2287	0.2148
	SVM	0.2319	0.2251	0.2284
	Our method	**0.3880**	**0.4652**	**0.4232**
Surprise	Naïve Bayesian	0.1765	0.0061	0.0119
	SVM	0.2038	0.0179	0.0329
	Our method	**0.3911**	**0.3261**	**0.3556**
Anger	Naïve Bayesian	0.1962	0.1071	0.1385
	SVM	0.2025	0.1673	0.1832
	Our method	**0.3854**	**0.3752**	**0.3802**
Disgust	Naïve Bayesian	0.2047	0.2664	0.2315
	SVM	0.2216	0.3079	0.2577
	Our method	**0.4543**	**0.5990**	**0.5167**
Fear	Naïve Bayesian	0.2009	0.2707	0.2306
	· SVM	0.1972	0.2253	0.2103
	Our method	**0.4778**	**0.3454**	**0.4010**
Sadness	Naïve Bayesian	0.2424	0.4008	0.3021
	SVM	0.2711	0.4223	0.3302
	Our method	**0.4845**	**0.4674**	**0.4758**
Average	Naïve Bayesian	0.2039	0.2133	0.1883
	SVM	0.2213	0.2276	0.2071
	Our method	**0.4318**	**0.4297**	**0.4254**

Wondered which factors play a vital role in the promotion, we conduct factor contribution analysis and several interesting sociological and psychological phenomena are discovered.

5.3   Factor Contribution Analysis

Herein, we test the contribution for each factor function used in the model. On the basis of the primitive model, we take out one of the factor and then examine the performance of the model while the other factors remain unchanged. The experimental results evaluated by F1-Measure are shown in Table 2.

As is shown in Table 3, the model which involves all factors achieves the best performance no matter it is evaluated by precision, recall or F1-Measure. The experimental results confirms that the $f3$, $f4$, $f5$ and $f6$ factor all make contribution to the promotion of the model. That is to say, on the image-based image social networks, the emotion influence indeed exist. It also proved that the correlations we observed in Section 3 is correct and effective.

More excitingly, it is clearly shown in the result that the $f3$ factor contributes the most in the model, which implies that whether the user is male or female and whether the user is single or taken did makes a great difference in the user's perception of images. It corresponds to the sociological and psychological theory. A.Fischer[10] pointed out that males tend to express more powerful emotions while females tend to express less powerful emotions, which indicates there is actually a gender difference in human emotion perception.

Besides, the $f6$ and $f4$ factor, which represents the intimacy between user and her friends and the size of her friends exert a profound influence on the user's emotional state. J.Pennebaker[11] discovered that people are more likely to share their emotion with those who are close to them, such as the family members, spouse and close friends.

J.Whitfield[13] found that happy people are more likely to be surrounded by happy people, while sad people tend to make friends with those who usually feel sad. People having similar emotions have a tendency to gather together, which illustrates the contribution of $f5$ factor.

Table 2. The contribution of different factor

Emotion	Model	Model-$f3$	Model-$f4$	Model-$f5$	Model-$f6$
Happiness	0.4232	0.4169	0.4136	0.4183	0.4149
Surprise	0.3556	0.3272	0.3208	0.3298	0.3303
Anger	0.3802	0.3686	0.3839	0.3835	0.3858
Disgust	0.5167	0.4459	0.4995	0.5022	0.5012
Fear	0.4010	0.3425	0.3519	0.3652	0.3642
Sadness	0.4758	0.4738	0.4643	0.4701	0.4692
Average	**0.4254**	**0.3958**	**0.4057**	**0.4115**	**0.4109**

6   Conclusion

In this paper, we study the problem of inferring emotions of social images leveraging influence analysis and proposed a novel method to solve the problem.

We first explored the patterns of emotion influence on the world's largest image-sharing website Flickr and observed several interesting psychological phenomena. Then we summarized these patterns into three types of correlations and introduced them in the factor graph model in the form of factor functions, which fulfills the modeling and inference of emotions of social images. By conducting experiments, we validate the effectiveness of our method and present a noteworthy promotion of the problem.

Image-based social networks such as Flickr and Instagram are thriving, making the problem of inferring emotions of social images of great significance. The research advances can provide a useful back-up for sociology and psychology and help Internet companies offer better services for customers.

Acknowledgement. This work is supported by the National Basic Research Program (973 Program) of China (2012CB316401), National Natural, and Science Foundation of China (61370023, 61003094). This work is partially supported by the major project of the National Social Science Foundation of China (13&ZD189) and funded by Microsoft Research Asia-Tsinghua University Joint Laboratory: FY14-RES-SPONSOR-111.

References

1. Tang, J., Zhang, Y., Sun, J., Rao, J., Yu, W., Chen, Y., Fong, A.: Quantitative study of individual emotional states in social networks. IEEE Transactions on Affective Computing 3(2), 132–144 (2012)
2. Machajdik, J., Hanbury, A.: Affective image classification using features inspired by psychology and art theory. In: Multimedia, pp. 83–92. ACM (2010)
3. Wang, X., Jia, J., Yin, J., Cai, L.: Interpretable aesthetic features for affective image classification. In: Proc. of ICIP 2013, Melbourne, pp. 3230–3234 (2013)
4. Hareli, S., Rafaeli, A.: Emotion cycles: On the social influence of emotion in organizations. In: Research in Organizational Behavior, vol. 28, pp. 35–59 (2008)
5. Fowler, J., Christakis, N., Steptoe, Roux, D.: Dynamic Spread of Happiness in a Large Social Network: Longitudinal Analysis of the Framingham Heart Study Social Network. British Medical Journal 338(7685), 23–27 (2009)
6. Ekman, P.: An argument for basic emotions. Cognition and Emotion 6(3-4), 169–200 (1992)
7. Fellbaum, C.: Wordnet. In: Theory and Applications of Ontology: Computer Applications, pp. 231–243 (2010)
8. Dong, Z., Dong, Q.: HowNet and the Computation of Meaning. World Scientific (2006)
9. Chang, C., Lin, C.: LIBSVM: A library for support vector machines. ACM TIST 2, 27: 1–27:27 (2011)
10. Fischer, A., Rodriguez Mosquera, P., van Vianen, A., Manstead, A.: Gender and culture differences in emotion. Emotion 4(1), 87–94 (2004)

11. Pennebaker, J., Zech, E., Rimé, B.: Disclosing and Sharing Emotion: Psychological, Social and Health Consequences. In: Stroebe, M.S., Stroebe, W., Hansson, R.O., Schut, H. (eds.) Handbook of Bereavement Research: Consequences, Coping, and Care, pp. 517–539. American Psychological Association, Washington, DC (2001)
12. Yang, Y., Jia, J., Zhang, S., Wu, B., Chen, Q., Li, J., Xing, C., Tang, J.: How Do Your Friends on Social Media Disclose Your Emotions? In: Proceedings of the 28th AAAI Conference on Artificial Intelligence
13. Whitfield, J.: The secret of happiness: Grinning on the Internet. Nature (June 26, 2008)
14. Borth, D., Ji, R., Chen, T., Breuel, T., Chang, S.: Large-scale visual sentiment ontology and detectors using adjective noun pairs. In: Multimedia, pp. 223–232. ACM (2013)
15. Shin, Y., Kim, E.: Affective prediction in photographic images using probabilistic affective model. In: ICIVR, pp. 390–397. ACM (2010)

Emotion Evolution under Entrainment in Social Media

Saike He[1], Xiaolong Zheng[1], Daniel Zeng[1,2], Bo Xu[3], Guanhua Tian[3],
and Hongwei Hao[3]

[1] State Key Laboratory of Management and Control for Complex Systems,
Institute of Automation, Chinese Academy of Sciences, Beijing 100190, China
[2] Department of Management Information Systems, University of Arizona,
Tucson, AZ 85721, USA
[3] Institute of Automation, Chinese Academy of Sciences, Beijing 100190, China
{saike.he,xiaolong.zheng,dajun.zeng,boxu,guanhua.tian,
hongwei.hao}@ia.ac.cn

Abstract. Emotion entrainment refers to the phenomenon that people gradually synchronize to other's emotion states through social interactions. Previous studies mainly focus on conducting laboratory experiments or small-scale offline surveys. Large-scale empirical studies on real-world emotion entrainment among individuals are still to be explored. Especially, determinants that influence this process are not clear. Also, how emotion evolves among people in a large scale population is still unknown. In this study, we attempt to conduct a large-scale empirical analysis on emotion entrainment based on online social media information. For this purpose, we develop a model-free framework to measure entrainment strength among people. Experimental results indicate that interaction partners with strong reciprocal entrainment tend to assume similar emotion states, and negative emotion is more empathetic in an intimate relationship. Especially, when the relationship is balanced, users are more emotionally similar to each other.

Keywords: Emotion Entrainment, Transfer Entropy, Social Media.

1 Introduction

The principle of emotion entrainment accounts for the convergence of people's rhythmic emotions through social interactions. For instance, when an individual feels unhappy, his friends may also feel unhappy, depending on the intimacy of their relationship. This principle is highly relevant to people's daily life, since it provides the basic currency in social relationships, and it is essential to the quality and scale of people's routine experience. Despite its importance, we have little understanding of the emotion dynamics in different kinds of relationships, as well as the determinants that steer the emotion trajectory during entrainment process. In addition, most previous investigation on entrainment only investigate through small scale or laboratory experiments, thus how emotion entrains over time in a large-scale, real-world setting remains a puzzle.

H.-Y. Huang et al. (Eds.): SMP 2014, CCIS 489, pp. 155–163, 2014.

In this paper, we attempt to explore the dynamics of emotion entrainment in a large scale setting. However, modeling emotion dynamics on a large scale at any detail is often challenging. Traditional bottom-up approaches are limited by their scalabilities due to high complexity in modeling, while more recent network based approaches are incapable to depict emotion dynamics accurately. As such, we propose to utilize a model-free approach, *transfer entropy*, to model emotion dynamics and derive a metric to quantify entrainment strength. Experiments on a dataset collected from Livejournal – a popular social media platform in the west – suggest that interaction partners with strong reciprocal entrainment tend to assume similar emotion states, and negative emotion is more empathetic in intimate relationship. When the relationship is balanced, the users are more emotionally similar to each other. Findings revealed in this paper may be useful for researchers and practitioners in understanding people's online relationships based on their emotion interactions, and informing whether their relationships are balanced or not.

The rest of the paper is structured as follows. Section 2 describes the dataset used and the metric for quantifying emotion entrainment. Section 3 reports the results of emotion evolution from three different perspectives. Section 4 concludes this paper with a summary and a discussion for future work.

2   Dataset and Metric

In this section, we first describe the dataset used, then proceed to present the technical details of modeling framework for emotion entrainment.

2.1   Dataset

Livejournal[1] is a blogging platform that allows its users to tag emotional labels when posting messages. Apart from system defined labels, Livejournal also allows users to define their own emotion labels. The dataset used in following studies (named as CHI06) is collected by Leshed and Kaye [1], and contains about 1.6 million bloggers who generate 18 million English blog posts. Besides, it provides a whole observation period from 1st May 2001 to 23rd Apr. 2005. This represents us with an entire emotion entrainment history.

Table 1. Categories of mood labels

Category	Examples of Mood Labels	Sample Number
Positive	great, elated, cheerful, ecstatic, jovial, fantastic, whee, triumphant, perky	285
Neutral	calm, so so, at peace, normal, ready for bed, working, thirsty, busy, blah, snuffly, warm	665
Negative	bored, sore, depressed, homicidal, crappy, yucky, remorseful, bitchy, befuddled, edgy	499
Total		1449

---

[1] http://www.livejournal.com/

For computation convenience, we select the most commonly used emotion labels (totally 1449) in the dataset, and group them into three categories: positive, neutral, and negative (Table 1). For the sake of mathematical manipulation, we assign labels in positive, neutral, and negative category with charges of [+1, 0, -1] respectively.

To facilitate analysis, in following studies, we randomly select 20,000 users who have posted more than 5 messages from the whole community.

2.2   Entrainment Metric

To quantify entrainment efficiently, we try to design our modeling framework with the fewest assumptions about exact social interactions. Also, we should distinguish entrainment directions as revealed by Will and Berg [2]. Consequently, we choose transfer entropy [3-6] as the metric to measure entrainment strength between each user pair. Specifically, if we record the emotion states of two users x and user y as two Markov processes $X = x_t$ and $Y = y_t$, then the entrainment strength from x to y can be defined as the transfer entropy from y to x, as shown in:

$$Entrain\ (X \to Y) = TE(Y \to X) = H\left(x_{t+1}\middle|\mathbf{x}_t^m\right) - H\left(x_{t+1}\middle|\mathbf{x}_t^m, \mathbf{y}_t^n\right) \tag{1}$$

where, $\mathbf{x}_t^m = \left(x_t, \ldots, x_{t-m+1}\right)$, $\mathbf{y}_t^n = \left(y_t, \ldots, y_{t-n+1}\right)$, while m and n are the orders of each of the Markov processes[2]; $H(*)$ calculates the entropy of the probability distribution enclosed.

This metric captures complex nonlinear emotion dynamics without modeling exact social interactions. In addition, it differentiates entrainment directions between dyadic interactions, which are often ignored by previous research. For detailed calculation of (1), please refer to Vicente $et\ al.$ [3].

3   Emotion Evolution under Entrainment

In this section, we primarily examine how emotion evolves driven by entrainment process. Then, we further our study by exploring how each type of emotion evolves through dyadic interactions. Finally, we explore how emotion entrains in different kinds of social relationship.

3.1   Emotion Evolution in General Case

Entrainment level reflects the social closeness among people [7], and specific kind of emotion are only shared between intimate relationships [8]. Thus, we hypothesize people bearing intimate relationship tend to be emotionally similar.

To test this hypothesis, we first define a variable to depict emotion disparity for a given user pair. For a user v, we represent his sequential emotion states at each time

---

[2] For simplicity, we hereafter take m=n=3.

stamp t within an observation period T as a vector $\mathbf{V} = \{e_1, \ldots, e_t, \ldots, e_T\}$. For user v_i and user v_j, their emotional difference is defined as the angle cosine of their emotion vector $\mathbf{V}_i$ and $\mathbf{V}_j$ respectively[3]:

$$\text{Emo\_Diff}\left(v_i, v_j\right) = \text{Ar}\cos\left(\mathbf{V}_i, \mathbf{V}_j\right) \tag{2}$$

where, Arcos() calculate the angle cosine between the two vectors enclosed. Emo_Diff spans with an interval of [0, 180], and higher value suggest larger emotion difference.

As both entrainment strength and emotion state are in constant flux, (2) provides a reliable measure for the disparity of emotion states for each user pair in a long enough observation period. The distribution of emotion difference over entrainment strength is shown in Fig. 1.

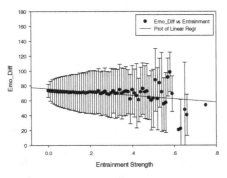

Fig. 1. Emotion difference over entrainment strength. Black node is the mean of Emo_Diff and error bars represent the standard deviation; blue line corresponds to the regression line ($y = b_0 + b_1 X$), where $b_1 = -22.197$.

Fig. 1 suggests that the Emo_Diff decreases as entrainment enhances (Pearson's correlation coefficient: -0.367, $p < 0.05$). This implies that user peers tend to be emotionally similar under strong reciprocal entrainment, which, in turn, suggests that emotion empathy is more efficient between intimate relationships.

3.2   Emotion Evolution Based on Type

Types of emotion shared between people are usually relation dependent, i.e. bad moods are only shared between intimate relationships [8]. Thus, we further our study by examining whether each type of emotion evolves differently through dyadic interactions. As such, we divide the selected users into three groups according to their average emotion charge, as suggested by an indicator function $C(v)$:

---

[3] We take t as 1 day, and T as 7 days in this study.

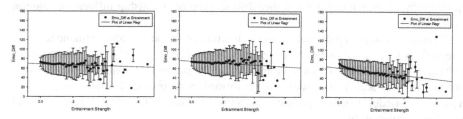

Fig. 2. Emotion evolution of each type. (a), (b), and (c) correspond to the evolution of positive, neutral, and negative emotion respectively; blue line corresponds the regression line $(y=b_0+b_1X)$, where b_1 in (a),(b), and (c) are -10.329, -18.894 and -38.234.

$$C(v) = \begin{cases} POS, & if \ 1/|M| \sum_{m \in M} Char(m) > \theta_{up}, \\ NEG, & if \ 1/|M| \sum_{m \in M} Char(m) < \theta_{low}, \\ NEU, & otherwise. \end{cases} \qquad (3)$$

where, M is a set of $|M| = N$ blog messages m written by user v; $Char(m)$ represents the emotion charge of message m; θ_{up} and θ_{low} are respectively the upper and lower charge boundary separating each user group. By assigning $\theta_{up0}=0$ and $\theta_{low} = -0.334$[4], we obtain 4261, 10625 and 5114 users in POS, NEU and NEG group respectively.

In separating selected users, we assume that users in each group generally experience a certain kind of 'emotion tone'. For instance, users in POS group overall experience positive moods in the whole observation period.

Under this grouping framework, we then examine how each type of emotion evolves within each user group, as shown in Fig. 2.

Fig. 2 indicates that for all the three types of emotion, their difference between user pairs decreases as entrainment strength enhances. This implies a noticeable correlation between emotion disparity and reciprocal entrainment strength. Pearson's correlation coefficients in POS, NEU, and NEG group are -0.138, -0.211, and -0.384 respectively ($p<0.05$). This result reveals that negative emotion appears to be more empathetic in intimate relationships. Also, this finding may explain why negative emotion is more influential to people, and why it spreads faster than other types of emotion in social media [9].

3.3   Emotion Evolution Based on Network Structure

Synchronization delay or phase shift is a common phenomenon in entrainment process [10, 11]. For instance, at the initial stage of emotional interactions, people's entrainment potential is high whereas their emotion states may be quite different. In such situations, the correlation between emotion difference and entrainment strength

---

[4] This corresponds to two times of the average emotion charge for all selected users, i.e. -0.167.

can be weak. However, this correlation tends to be strong in more stable relationships. In this subsection, we explore whether distinct type of friendship influences emotion evolution.

Previous research argues that network structure of friendship also influences social interactions [12, 13]. Among these, social balance theory [14, 15] examines triads of individuals in social networks, and argues that people in an unbalanced relationship are incentive to adjust their social states toward more balanced formation [16, 17]. Thus, we wonder whether people's emotion entrains differently in balanced and unbalanced relationship.

To study this issue, we first redefine a pairwise binary relationship in a triad as:

$$E(v_i, v_j) = \begin{cases} +, & if\ \min\left(Entrain\left(v_i \rightarrow v_j\right), Entrain\left(v_i \leftarrow v_j\right)\right) \geq \beta, \\ -, & if\ \max\left(Entrain\left(v_i \rightarrow v_j\right), Entrain\left(v_i \leftarrow v_j\right)\right) < \beta, \\ NULL, & otherwise. \end{cases} \tag{4}$$

where, $E(v_i, v_j)$ denotes the polarity of the singed relationship (+ or -). min(), and max() calculate the minimum and maximum value of the two arguments enclosed. β is the threshold that distinguishes positive and negative relationships[5].

Given (4), the constructed triads in balanced and unbalanced formation are shown in Fig. 3. According to the social balance theory, a triad is said to be balanced if the algebraic multiplication of signs in the triad relation has a positive value, or unbalanced otherwise.

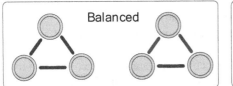
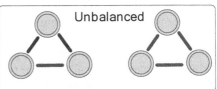

Fig. 3. Social Balance. Blue edge and red edge correspond to positive (+) and negative (-) relationship respectively. The two left triads are balanced, while the two right triads are unbalanced.

We then make a hypothesis that entrainment procedure enhances social ties. Under this assumption, social networks should become gradually balanced as entrainment process proceeds. To test this hypothesis, we introduce the global balance index [16] to measure social balance for a given social network, as defined below:

$$BI = \frac{\sum\limits_{J \leq I} T_{balanced}}{\sum\limits_{I} T_{total}} \tag{5}$$

---

[5] Without loss of generality, we set β as the average entrainment strength.

where, $T_{balanced}$ denotes the number of balanced triads, T_{total} denotes the total number of triads in the whole networks, J and I represent the number of balanced and whole triads respectively.

We now investigate how the balance level changes along with proceeding of entrainment. To this end, for every seven days, we calculate the entrainment relationship for all selected users, as shown in Fig. 4.

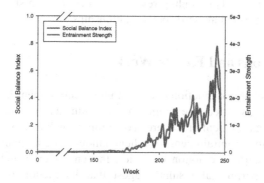

Fig. 4. Evolution of social balance. Green curve corresponds to the social balance index, and the red curve corresponds to the evolution of entrainment strength.

In Fig. 4, the balance level of the community on Livejournal increases as the entrainment strength develops. This finding verifies our previous hypothesis about entrainment and social balance. In addition, it is also consistent with Heider's conclusion that every social network tends to achieve higher balance level [14].

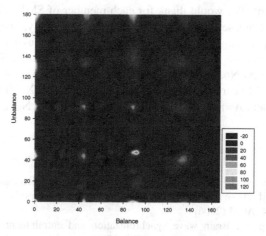

Fig. 5. Emotion difference in balanced and unbalanced friendship

Further, for each specific user pair, we explore how their emotion changes in balanced and unbalanced structure, and whether their emotions develop more similar to each other in balanced social relationship.

To clarify this issue, for each user pair, we plot the EMO_Diff in balanced relationships over that in unbalanced relationships (Fig. 5).

In Fig. 5, we notice the average Emo_Diff in a balanced relationship (average value: 60.065) is smaller than that in unbalanced one (average value:73.218). This difference is statistically significant with a p-value smaller than 0.001 according to a paired two-tailed t-test. This finding reveals that the emotion states are more similar within user pairs bearing balanced social relationships.

4   Conclusions and Future Work

In this paper, we explored emotion evolution under entrainment in a large-scale, real-world setting. We also tried to examine the determinants that may steer the emotion trajectory during entrainment procedure. To facilitate analytics, we proposed a model-free metric to quantify entrainment for a huge amount of user pairs. With this metric, we attempted to explore emotion evolution in general case and for each specific emotion type. Experimental results suggest that interaction partners with strong reciprocal entrainment tend to assume similar emotion states, and negative emotion is more empathetic in an intimate relationship. Besides, emotion evolution is also relation dependent, analysis based on social balance theory reveals that users are more emotionally similar to each other in balanced relationships.

In our future study, we hope to develop an interaction model to capture the emotion dynamics, and then try to give theoretical explanations for findings discovered in this paper.

Acknowledgement. We would thank for each member of SMILES group in Institute of Automation, Chinese Academy of Sciences. Especially, we would thank for Kainan Cui, Zhu Zhang, and Chuan Luo for useful discussions. This work was supported in part by the following grants: The National Natural Science Foundation of China under Grant No. 71103180, 71025001, 71472175, 61175040, 61303172, and 91124002; by the Ministry of Health under Grant No. 2012ZX10004801, by the Early Career Development Award of SKLMCCS, and by the Grant No. 2013A127.

References

1. Leshed, G., Kaye, J.J.: Understanding how bloggers feel: Recognizing affect in blog posts. In: CHI 2006 Extended Abstracts on Human Factors in Computing Systems, pp. 1019–1024. ACM (2006)
2. Will, U., Berg, E.: Brain wave synchronization and entrainment to periodic acoustic stimuli. Neuroscience Letters 424, 55–60 (2007)

3. Vicente, R., Wibral, M., Lindner, M., Pipa, G.: Transfer entropy—a model-free measure of effective connectivity for the neurosciences. Journal of Computational Neuroscience 30, 45–67 (2011)

4. He, S., Zheng, X., Zeng, D., Cui, K., Zhang, Z., Luo, C.: Identifying peer influence in online social networks using transfer entropy. In: Wang, G.A., Zheng, X., Chau, M., Chen, H. (eds.) PAISI 2013. LNCS, vol. 8039, pp. 47–61. Springer, Heidelberg (2013)

5. He, S., Bao, X., Ma, H., Zheng, X., Zeng, D., Xu, B., Li, C., Hao, H.: Characterizing Emotion Entrainment in Social Media. In: The 2014 IEEE/ACM International Conference on Advances in Social Network Analysis and Mining. ACM (2014)

6. He, S., Zheng, X., Zeng, D., Xu, B., Li, C., Hao, H.: Ranking Online Memes in Emergency Events Based on Transfer Entropy. In: IEEE Joint Intelligence and Security Informatics Conference (JISIC), The Hague, The Netherlands (2014)

7. Harrison, D.A., Mohammed, S., McGrath, J.E., Florey, A.T., Vanderstoep, S.W.: Time matters in team performance: Effects of member familiarity, entrainment, and task discontinuity on speed and quality. Personnel Psychology 56, 633–669 (2003)

8. Tang, J., Zhang, Y., Sun, J., Rao, J., Yu, W., Chen, Y., Fong, A.C.M.: Quantitative study of individual emotional states in social networks. AIEEE Transactions on Affective Computing 3, 132–144 (2012)

9. Fan, R., Zhao, J., Chen, Y., Xu, K.: Anger is more influential than joy: Sentiment correlation in weibo. arXiv preprint arXiv:1309.2402 (2013)

10. Rensing, L., Ruoff, P.: Temperature effect on entrainment, phase shifting, and amplitude of circadian clocks and its molecular bases. Chronobiology International 19, 807–864 (2002)

11. Aschoff, J., Hoffmann, K., Pohl, H., Wever, R.: Re-entrainment of circadian rhythms after phase-shifts of the Zeitgeber. Chronobiologia 2, 23–78 (1974)

12. Ugander, J., Backstrom, L., Marlow, C., Kleinberg, J.: Structural diversity in social contagion. Proceedings of the National Academy of Sciences 109, 5962–5966 (2012)

13. Zheng, X., Zhong, Y., Zeng, D., Wang, F.-Y.: Social influence and spread dynamics in social networks. Frontiers of Computer Science 6, 611–620 (2012)

14. Heider, F.: Attitudes and cognitive organization. The Journal of Psychology 21, 107–112 (1946)

15. Zheng, X., Zeng, D., Wang, F.-Y.: Social balance in signed networks. Information Systems Frontiers, 1–19 (2014)

16. Khanafiah, D., Situngkir, H.: Social balance theory: Revisiting Heider's balance theory for many agents (2004)

17. Traag, V.A., Van Dooren, P., De Leenheer, P.: Dynamical models explaining social balance and evolution of cooperation. PloS One 8, e60063 (2013)

Topic Related Opinion Integration for Users of Social Media

Songxian Xie, Jintao Tang, and Ting Wang

School of Computer Science, National University of Defense Technology, Hunan, PRC
{xsongx,tangjintao,tingwang}@nudt.edu.cn

Abstract. Social media such as Twitter, has become a valuable source for mining opinions of users about all kinds of topics. In this paper, we investigate how to automatically integrate topic related opinions expressed by a user in User-Generated Content (UGC). We propose a general subjectivity model by combining topics and fine-grained opinions towards each topic, and design an efficient algorithm to establish the model. We demonstrate utility of our model in the opinion prediction problem and verify the effectiveness of our model qualitatively and quantitatively in a series of experiments on real Twitter data. Results show that the proposed model is effective and can generate consistent integrated opinion summaries for users. Furthermore, the proposed model is more suitable for social media context, thus can reach better performance in an opinion prediction task.

Keywords: LDA, social media, opinion integration, subjectivity model.

1 Introduction

With the rise of content-based social media such as Twitter, millions of users are more and more willing to publish online short messages to express their opinions on a great variety of topics they are interested in. The wide coverage of topics, dynamics of discussion, and abundance of opinions imbedded in the social media data make them extremely valuable source for mining users' opinions about all kinds of topics (e.g., products, political figures, etc.), which in turn can enable a wide range of applications, such as opinion search for ordinary users, opinion tracking for business intelligence, and user behavior prediction for targeted advertising. However, with such a large scale of information source, it is quite challenging to integrate and digest all the opinions from different users. For example, a query "iPhone" on Twitter (as of Jan. 14, 2014) returns 830,879 tweets of 231,233 users, suggesting that there are many users have expressed opinions more than once about iPhone in their tweets. To enable an application to benefit from all kinds of opinions of different users, it is thus necessary to automatically integrate and present an overall opinion summary for each user [1]. In fact, users often publish several messages on the topics they are interested in, therefore how to find these topics and integrate opinions towards each topic scattered in many independent tweets of a user poses special challenge for opinion mining related researchers.

H.-Y. Huang et al. (Eds.): SMP 2014, CCIS 489, pp. 164–174, 2014.
© Springer-Verlag Berlin Heidelberg 2014

In this paper, we propose a combining model (named as subjectivity model) by incorporating topics and opinions at the user level, of which one part represents topics of interest distribution, while the other part represents the distribution of opinions towards these topics. Specifically, we propose a general method to solve this integration problem in three steps illustrated as in Figure 1: (1) extract topics of interest from tweets of a user using user-level LDA; (2) extract separate opinion and topic for each tweet with sentiment and topic analysis (3) summarize and integrate the extracted opinions towards each topic to form a subjectivity model for each user.

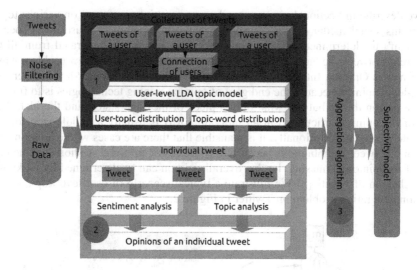

Fig. 1. Framework

The rest of the paper is organized as follows. In Section 2, we introduce related works, and we formally define the novel problem of opinion integration in Section 3. After that, we present our model and analyze the difference with generative model in Section 4. We discuss our experiments and results in Section 5. Finally, we conclude our work and point out future target.

2 Related Works

Sentiment analysis is a popular research area and previous researches have mainly focused on reviews or news comments [2, 3]. Recently, there have been many works on sentiment analysis on Twitter, mainly focusing on the tweet level [4, 5, 6, 7, 8], of which, the techniques employed are generally standard tweet-level algorithms that ignore many special characteristics of social media. There have been also some previous works on automatically determining user-level opinions or ideology [9, 10], generally looking at information embedded in the contents that the users generate. Most of related researches mainly focused on identification of sentimental object [11], or detection of objects' sentimental polarity [12] without considering the topic aspects.

Since the introduction of topic model such as LDA [13], various extended models have been used for topic extraction from large-scale corpora at user level [14, 15]. Topic models can also be utilized in sentiment analysis to correlate sentiment with topics. Mei et al. [16] and Lin et al. [17] incorporated topic models and sentiment analysis for reviews and blogs.

3   Opinion Integration Problem

As we describe in Section 1, a user usually posts multiple messages on various topics during his social media usage. Therefore what's the opinion of a user on a specific topic can't be determined from just one tweet, but should be integrated from all the topic related tweets he has posted. In this paper, we put forward a new problem which is defined as Opinion Integration Problem (OIP). We focus on user-level rather than tweet-level opinion because the end goal of opinion mining technologies is to find out what a person thinks but not what only a piece of message states, and the identification of the opinion articulated in an individual text is usually a middle step for that ultimate objective. Additionally, it is plausible that there are cases where opinions of a user in one tweet is ambiguous because they are restricted to be so short that the context of its opinion is missing, but his overall opinion can be determined by looking at his collection of tweets [8].We illustrate a typical scenario of user-level topic related opinion integration problem on Twitter in Figure 2.

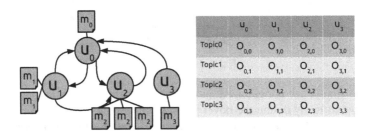

Fig. 2. Illustration of Opinion Integration Problem

Definition 1(Opinion Integration Problem). *As shown in the figure, there is a heterogeneous network of Twitter consisted of users set* $V = \{u_i\}$, *directional relations set* $E = \{(u_i, u_j) | u_i, u_j \in V\}$ *of all users, and their associated tweets* $M_i = \{m_i\}$, *in which topics (denoted as* $T = \{Topic_k\}$ *) and opinions of tweets can be determined and extracted. For a user* u_i, *his opinion (denoted as* $O_{i,k}$ *) towards topic* $Topic_k$ *is not the opinion imbedded in his single tweet* m_i, *but the integrated opinion from all his tweets* $M_i = \{m_i\}$.

There are two important factors that must be taken into considerations for the OIP problem. Firstly, topics both users and tweets talk about should be determined in a same topic space so as the target of opinion is consistent. Secondly but most importantly, opinions and topics are closely related, tweets of a user around some topic often cover a mixture of aspects related to that topic with different preferences. Different opinions may be expressed by the user towards different aspects, where users may like one aspect of a topic but dislike others. Therefore, how to integrate all opinions of tweets related to a topic into one holistic opinion and represent it reasonably poses special challenge. In this paper, we propose a novel subjectivity model to meet these two challenges.

4 Subjectivity Model

In this section, we give a formal definition of the model we work with to meet the challenges of OIP problem, which has been substantially defined and described in our previous work [18]. Here we only repeat the definition and the algorithm of model establishment, for more details, please refer to our paper [18]. Usually user level opinion is to classify each user's sentiment on a specific topic into one of two polarities: "Positive" and "Negative". "Positive" means that the user supports or likes the target topic, whereas "Negative" stands for the opposite. However in our model we adopt a broad "opinion" definition as sentiment coverage towards a topic over a fine-grained sentiment values to differentiate subtle opinions of users. For example one is more positive about a topic with sentiment strength 8 than another user with sentiment strength 7. At the same time, we define opinion of a user as a probabilistic distribution over the sentiment values instead of one single value, considering the user may express his different opinion on different aspects of the same topic. The notion of "opinion" is quite vague; we adopt this broad definition to ensure generality of the model. We frame the model in the context of Twitter to keep things concrete, although adaptation of our model to other social network settings is straightforward. We name our model as "subjectivity model" as it models the subjective information in the content generated by a user. Therefore, we give a formal definition of the subjectivity model under the context of Twitter as follows.

4.1 Definition

Let $G = (V, E)$ denotes a social network on Twitter, where V is a set of users, and $E = V \times V$ is a set of follow relationships between users. For each user $u \in V$, there is a tweets collection M_u denoting his message history. We assume that there is a topic space T containing all topics users in V talk about, and a sentiment space S to evaluate their opinions towards these topics. For the "**subjectivity**" of user $u \in V$, we refer to both topics and opinions articulated in his tweets collection M_u.

Definition 2 (Subjectivity Model). *The subjectivity model $P(u)$ of user u, is the combination of topics $\{t\}$ the user talks about in topic space T and his opinions O_t towards each topic distributed over sentiment space S.*

$$P(u) = \{(t, w_u(t), \{d_{u,t}(s) \mid s \in S\}) \mid t \in T\} \tag{1}$$

where:

- *with respect to user u, for each topic $t \in T$, its weight $w_u(t)$ represents the distribution of the user's interests on it, subject to $\sum_{t=1}^{|T|} w_u(t) = 1$.*
- *opinion of the user towards topic t is modeled as a topic related sentiment distribution over sentiment space S, $O_t = \{d_{u,t}(s) \mid s \in S\}$, subject to $\sum_{s=1}^{|S|} d_{u,t}(s) = 1$.*

Subjectivity model aims at obtaining the topic related refined sentiment for investigating user-level opinion mining, which can get a comprehensive understanding of the subjectivity for a user by modeling both his topics of interest and opinions towards each topic.

4.2   Establishment of Subjectivity Model

According to the definition of subjectivity model, there are two distributions to model the subjectivity: the topic distribution and the opinion distribution for each topic. Both of them need to be inferred from historic content produced by users.

For users set V of a social network, we denote tweets set published by a user $u \in V$ as $M_u = \{m_u\}$. M_u is concatenated to a document d_u to construct topic space $T = \{t_i \mid i = 1, \cdots K\}$ with user-level LDA model. The topic model is built with parameter θ representing the distribution of each user over topics in the topic space T, and parameter β representing the distribution of each topic over the vocabulary of all tweets. SentiStrength [25] is applied to each tweet m in collection M_u and outputs sentiment strength s_m for tweet m. With statistical topic analysis and opinion analysis for each user and tweet, we put forward a novel algorithm to concrete subjectivity model $P(u)$ for user u as algorithm 1. In the algorithm, we assume the sentiment of tweet m is related to every topic it talks about in Z_m for simplicity.

Algorithm 1. Establishment of subjectivity model.

Input: The users set of asocial network V ;
       The tweets set published by each user u, M_u ;

Output: The subjectivity model for each user u, $P(u)$;

Topic analysis with a user-level LDA, getting a topic model $P(\theta, \beta \mid M_u, V)$;

for all tweet $m \in M_u$ **do**

    Sentiment analysis, outputting sentiment of m, s_m ;

end for

for user $u \in V$ **do**

    the topic distribution is the corresponding component of parameter θ, θ_u ;

    the topics u tweets about are $Z_u = \{t \mid p(t \mid \theta_u) > 0, t \in T\}$;

---

end for

for $m \in M_u$ **do**

    topics of m can be identified by the topic model:

$$Z_m = \{t \mid p(t \mid \theta, \beta, Z_u) > 0, t \in T\} \tag{2}$$

end for

for each topic $t \in Z_u$ **do**

    for sentiment value $s \in S$ **do**

        count the number of tweets that talk about topic t with sentiment value s:

$$N_s = \sum_{m \in M_u \wedge s_m = s \wedge t \in Z_m} I(s_m) \tag{3}$$

    end for

    calculating opinion towards topic t:

$$O_t = \left\{ \frac{N_s}{\sum_{s \in S} N_s} \mid s \in [0, S] \right\} \tag{4}$$

end for

establishing subjectivity model of user u:

$$P(u) = \left\{ (t, p(t \mid \theta_u), \left\{ \frac{N_s}{\sum_{s \in S} N_s} \right\}) \mid t \in Z_u, s \in S \right\} \tag{5}$$

return $P(u)$.

4.3 Application of Subjectivity Model

The learned subjectivity model can be used to help with many applications such as opinion mining and behavior prediction (retweet, follow, etc.). Here we demonstrate one application on, i.e., how the learned model can help improve the performance of user opinion prediction. Our strategy is based on the premises that users usually tend to express their opinions consistently. In other words, positive and negative opinions are not randomly expressed by people. E.g., a user who supports a candidate in an election will tend to post positive tweets on a regular basis. Technically, social theories say that the user exhibits a varying degree of bias, which is his subjectivity [19].

We formulate the opinion prediction of a user as a triplet in the form of $< author, m, t >$, where author is the user who post tweet m, which talks about topic t. The goal is to predict the polarity $p = \{positive, negative\}$ of tweet m toward topic t. For such a problem, the dominant approach relies on extracting textual patterns from the tweet m and exploiting these patterns to predict its polarity.

However subjectivity model of a user provides information that is more robust to a single tweet short of context, as it is more consistent than typical textual information.

Thus, we propose an alternative approach to improve the performance of opinion mining of a single tweet based on subjectivity model of its author. Specifically, for tweet m, subjectivity model of its author $P(author)$ can be concreted according to algorithm 1. Let s_m denote its sentiment value calculated with some sentiment classifier such as SentiStrength. The topic tweet m talks about can be identified with equation 2 in algorithm 1:

$$\hat{t} = \arg\max(\hat{P}(t \mid \theta, \beta, Z_u) \mid t) \tag{6}$$

Thus opinion distribution of the user author can be identified from his subjectivity model $P(author)$: $O_{author,\hat{t}}$, which is a distribution over sentiment value space S. We can get a normalized sentiment value of the user on topic $\hat{t}$:

$$\hat{s}_m = \sum_{i \in T} d_i * v_i \tag{7}$$

where v_i denotes the sentiment value and d_i denotes the corresponding dimension of the sentiment distribution. Now we can predict the polarity p by smoothing the sentiment of tweet m with the normalized sentiment value of its author:

$$p = \begin{cases} positive & if \quad \dfrac{\hat{s}_m + s_m}{2} > \dfrac{|S|}{2} + 1; \\ negative & if \quad \dfrac{\hat{s}_m + s_m}{2} < \dfrac{|S|}{2}; \\ neutral & otherwise. \end{cases} \tag{8}$$

5   Experiment

5.1   Dataset and Settings

We use an off-the-shelf dataset [20], which is crawled from Twitter through its open API. The details about the dataset can be summarized as Table 1.

Table 1. Twitter Dataset Statistics

Total users 139,180	Friends per user 14.8
Total edges 4,175,405	Followers per user 14.9
Total tweets 76,409,820	Tweets per user 549

It is time-consuming to establish subjectivity model with the 139,180 users directly for the computational complexity of LDA. However, the principle of homophily [21], or "birds of a feather flock together" [22] suggests that users that are "connected" closely may tend to talk about similar topics and hold similar opinions [23]. On Twitter, the connections a user creates may correspond to approval or a desire to pay attention, or suggestive of the possibility of common topics and opinions. Therefore we adopt the community structure of social network to divide the 139,180 users into different community and establish subjectivity model for a user in his community local network. The communities are found with the packages igraph[1]. There are 106 communities in the global network, and 73 communities consist of users less than 15, for which topics can't be found effectively with LDA, so we filter out users in these communities. At the same time, we also filter out 15,756 users who are inactive with tweets less than 5, only tweet themselves with words less than 3, or only publish content with url links. In the final dataset, there are 122,329 users distributed in 33 communities. The subjectivity model for each user is established within his own community as algorithm 1.

Besides our model, we also conduct a set of experiments comparing with other topic-sentiment model including JST and TSM. The symmetry Dirichlet priors of topic models were set to 50/T and 0.01 respectively. The asymmetry sentiment prior empirically was set to (0.01, 1.8) for JST. All results were averaged over 5 runs with 2000 Gibbs sampling iterations.

5.2   Case Study

In order to qualitatively evaluate the effectiveness of our method, we give a vivid example of a user's subjectivity model, who has published 533 tweets. All his tweets are illustrated as Figure 3(a) in a word-cloud figure.

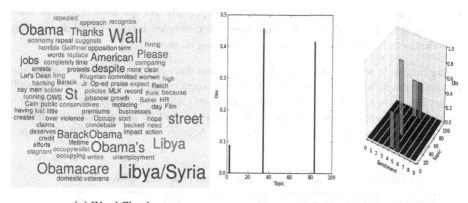

(a) Word Cloud.   (b) Subjectivity Model.

Fig. 3. Example of an user. In the subjectivity model, left sub-graph denotes interests distribution on topic 2, 32 and 83: $(w_u (2) = 0.08, w_u (32) = 0.48, w_u (83) = 0 \triangleright 44)$. The right sub-graph denotes opinions towards topics: $O_2 = (d_{u,2} (4) = 0.5, d_{u,2} (5) = 0.5)$, $O_{32} = (d_{u,32} (4) = 1 \triangleright 0)$, $O_{83} = (d_{u,83} (4) = 0.5, d_{u,83} (5) = 0.5)$.

Figure 3(b) is the visualized subjectivity model of the user in a [0, 100] topic space and a [0, 8] sentiment space, which is established according to our method. It is obvious that the user is interested in three topics (topic 2: "#Obamacare", topic 32:

---

[1] http://igraph.org/

"#libya" and topic 83:"#occupywallst"), and the left part of Figure 3(b) denotes the weights of his topics of interest. The right part denotes the opinions of the user towards three topics, in which he is neutral to topic "#libya" with 100% distribution on sentiment strength value 4, positive to topic "#Obamacare" and "#occupywallst" with 50% on value 4 and 50% on value 5. From the example, it is demonstrated that our model can give a detail description for the subjectivity of users in that it can model not only the interest distribution but also opinion coverage over a fine-grained sentiment.

5.3   Opinion Prediction Performance

To directly evaluate the effectiveness of our model quantitatively, we compare our model with other two generative topic-sentiment model (TSM and JST) with the number of topic is set to 50, 100, 150 and 200 iteratively. Short of labeled training data, we only compare our method with three state-of-the-art unsupervised sentiment analysis methods in the performance of opinion prediction.

- OF: OpinionFinder is a publicly available software package for sentiment analysis that can be applied to determine sentence-level subjectivity, i.e. to identify the emotional polarity (positive or negative) of sentences [24].
- S140: Sentiment140 can automatically classifying the sentiment of tweets using distant supervision with training data consisted of Twitter messages with emoticons.
- STR: SentiStrength package has been built especially to cope with sentiment analysis in short informal text of social media. It combines lexicon-based approaches with sophisticated linguistic rules adapted to social media [25].

We randomly select 1,000 target users from our dataset with at least 80 tweets, and select one random tweet for each user from his tweets collection to form a set of 1,000 tweets for evaluation. In order to identify topic of each tweet easily, the tweets with hashtag are prior to be selected. All 1,000 tweets in the test set are manually labeled with sentiment polarity as the golden standard. Accuracy is used as our performance measurement, and the result is list in Table 2.

Table 2. Accuracy performance. A significant improvement over OF with*

Method	50	100	150	200
OF	65.85%			
S140	70.45%*			
STR	69.98%*			
TSM	63.46%	72.94%*	67.83%	66.65%
JST	61.25%	68.57%*	75.88%*	67.03%
SUB	71.53%*	81.05%*	78.32%*	74.54%*

As can be observed from the result table that:

Firstly, the performance of OpinioFinder is the lowest with 65.85% accuracy, and we think the reason lying in that it is designed for the review and not adapts to tweets with informal language usage;

Secondly, other two unsupervised sentiment methods (Sentiment140: 70.45% and SentiStrength: 69.98%) outperform OpinioFinder significantly.

Thirdly, overall, the two generative models outperform OpinionFinder significantly, which demonstrates the importance of relating sentiment to the topics of users. Their performances are a little better than Sentiment140 and SentiStrength, but not significantly.

Finally, our method (SUB) outperforms all three unsupervised sentiment methods significantly with all four topic settings, and improves the performance of Senti-Strength significantly by combining subjectivity model of users with content of tweet. Compared with two generative models, our model outperforms TSM significantly, and gets a little better performance than JST. We think it is because sentiment analysis technique of our model is more suitable for the Twitter language, for it can extract subtle sentiment imbedded in special language characteristics such as repeated letters and emoticons.

6   Conclusion

In this paper, we define and investigate a novel opinion integration problem for social media users. We propose a subjectivity model to solve this problem in a three-stage framework and design an algorithm to establish the subjectivity model from historical tweets of users. With this model, we can automatically generate an integrated opinion summary that consists of both topics of interest distribution and topic related opinion distribution for a user. The proposed model is demonstrated effective in the application of opinion prediction. Experiments on Twitter data show that the proposed model can effectively describe topic related opinions with two probabilistic distributions and clearly outperforms generative models in the opinion prediction task. In the future, we will apply our model in several social network analysis applications to testify its effectiveness.

Acknowledgments. The research is supported by the National Natural Science Foundation of China (Grant No. 61170156 and 61202337).

References

1. Lu, Y., Zhai, C.: Opinion integration through semi-supervised topic modeling. In: Proceedings of the 17th International Conference on World Wide Web, pp. 121–130. ACM (2008)
2. Pang, B., Lee, L.: Opinion mining and sentiment analysis. Foundations and Trends in Information Retrieval 2(1-2), 1–135 (2008)
3. Liu, B.: Sentiment analysis and opinion mining. Synthesis Lectures on Human Language Technologies 5(1), 1–167 (2012)
4. Barbosa, L., Feng, J.: Robust sentiment detection on twitter from biased and noisy data. In: Proceedings of the 23rd International Conference on Computational Linguistics: Posters, pp. 36–44. Association for Computational Linguistics (2010)
5. Davidov, D., Tsur, O., Rappoport, A.: Enhanced sentiment learning using twitter hashtags and smileys. In: Proceedings of the 23rd International Conference on Computational Linguistics: Posters, pp. 241–249. Association for Computational Linguistics (2010)
6. Jiang, L., Yu, M., Zhou, M., Liu, X., Zhao, T.: Target-dependent twitter sentiment classification. In: Proceedings of the 49th Annual Meeting of the Association for Computational Linguistics: Human Language Technologies, vol. 1, pp. 151–160. Association for Computational Linguistics (2011)

7. Li, G., Hoi, S.C., Chang, K., Jain, R.: Micro-blogging sentiment detection by collaborative online learning. In: 2010 IEEE 10th International Conference on Data Mining (ICDM), pp. 893–898. IEEE (2010)

8. Tan, C., Lee, L., Tang, J., Jiang, L., Zhou, M., Li, P.: User-level sentiment analysis incorporating social networks. In: Proceedings of the 17th ACM SIGKDD International Conference on Knowledge Discovery and Data Mining, pp. 1397–1405. ACM (2011)

9. Mostafa, M.M.: More than words: Social networks text mining for consumer brand sentiments. Expert Systems with Applications 40(10), 4241–4251 (2013)

10. Malouf, R., Mullen, T.: Taking sides: User classification for informal online political discourse. Internet Research 18(2), 177–190 (2008)

11. Liu, H., Zhao, Y., Qin, B., Liu, T.: Comment target extraction and sentiment classification. Journal of Chinese Information Processing 24(1), 84–89 (2010)

12. Zhai, Z., Liu, B., Xu, H., Jia, P.: Constrained LDA for grouping product features in opinion mining. In: Huang, J.Z., Cao, L., Srivastava, J. (eds.) PAKDD 2011, Part I. LNCS (LNAI), vol. 6634, pp. 448–459. Springer, Heidelberg (2011)

13. Blei, D.M., Ng, A.Y., Jordan, M.I.: Latent dirichlet allocation. The Journal of Machine Learning Research 3, 993–1022 (2003)

14. Rosen-Zvi, M., Griffiths, T., Steyvers, M., Smyth, P.: The author-topic model for authors and documents. In: Proceedings of the 20th Conference on Uncertainty in Artificial Intelligence, pp. 487–494. AUAI Press (2004)

15. Ramage, D., Hall, D., Nallapati, R., Manning, C.D.: Labeled lda: A supervised topic model for credit attribution in multi-labeled corpora. In: Proceedings of the 2009 Conference on Empirical Methods in Natural Language Processing, vol. 1, pp. 248–256. Association for Computational Linguistics (2009)

16. Mei, Q., Ling, X., Wondra, M., Su, H., Zhai, C.: Topic sentiment mixture: modeling facets and opinions in weblogs. In: Proceedings of the 16th International Conference on World Wide Web, pp. 171–180. ACM (2007)

17. Lin, C., He, Y.: Joint sentiment/topic model for sentiment analysis. In: Proceedings of the 18th ACM Conference on Information and Knowledge Management, pp. 375–384. ACM (2009)

18. Xie, S., Tang, J., Wang, T.: Resonance elicits diffusion: Modeling subjectivity for retweeting behavior analysis. Cognitive Computation, 1–13 (2014)

19. Walton, D.N.: Bias, critical doubt and fallacies. Argumentation and Advocacy 28, 1–22 (1991)

20. Li, R., Wang, S., Deng, H., Wang, R., Chang, K.C.C.: Towards social user profiling: unified and discriminative influence model for inferring home locations. In: KDD, pp. 1023–1031 (2012)

21. Lazarsfeld, P.F., Merton, R.K.: Friendship as a social process: A substantive and methodological analysis. In: Berger, M., Abel, T. (eds.) Freedom and Control in Modern Society. Van Nostrand, New York (1954)

22. McPherson, M., Smith-Lovin, L., Cook, J.M.: Birds of a feather: Homophily in social networks. Annual Review of Sociology, 415–444 (2001)

23. Thelwall, M.: Emotion homophily in social network site messages. First Monday 15(4) (2010)

24. Wilson, T., Wiebe, J., Hoffmann, P.: Recognizing contextual polarity in phrase-level sentiment analysis. In: Proceedings of the Conference on Human Language Technology and Empirical Methods in Natural Language Processing, pp. 347–354. Association for Computational Linguistics (2005)

25. Thelwall, M., Buckley, K., Paltoglou, G., Cai, D., Kappas, A.: Sentiment strength detection in short informal text. Journal of the American Society for Information Science and Technology 61(12), 2544–2558 (2010)

Relationship between Reviews Polarities, Helpfulness, Stars and Sales Rankings of Products: A Case Study in Books

Qingqing Zhou[1,2] and Chengzhi Zhang[1,3,*]

[1] Department of Information Management, Nanjing University of Science and Technology, Nanjing, China
[2] Alibaba Research Center for Complex Sciences, Hangzhou Normal University, Hangzhou, China
[3] Jiangsu Key Laboratory of Data Engineering and Knowledge Service (Nanjing University), Nanjing, China
breeze7zhou@gmail.com, zhangcz@njust.edu.cn

Abstract. To help customers, especially the customers without explicit purchasing motivation, to obtain valuable information of products via E-commerce websites, it is useful to predict sales rankings of the products. This paper focuses on this problem by finding relationship between reviews, star level and sales rankings of products. We combine various factors with the information of helpfulness and conducting correlation analysis between sales rankings and our combinations to find the most correlative combinations, namely the optimal combinations. We use three domains of books from Amazon.cn to conduct experiments. The main findings show that helpfulness is really useful to predict book sales rankings. Different domains of books have different optimal combinations. In addition, in consideration of helpfulness, the combination of number of positive reviews, score of review stars and score of frequent aspects is the most correlative combination. In this paper, although reviews on Amazon.cn are written in Chinese, our method is language independent.

Keywords: "sales ranking", "sentiment analysis", "helpfulness of reviews".

1 Introduction

With the development of E-commerce, people are more likely to buy products online. It is time-consuming for customers to understand products more deeply and choose the favorite ones. For customers who have no explicit purchasing motivation, sales ranking is a good index for their choices. However, there only partly products are in the list of sales ranking, new products are always out of it.

Many existing researches use reviews of products to predict sales rankings, few of them combine reviews and stars of the products together [1]. In this paper, we present

---

*Corresponding author.

H.-Y. Huang et al. (Eds.): SMP 2014, CCIS 489, pp. 175–186, 2014.

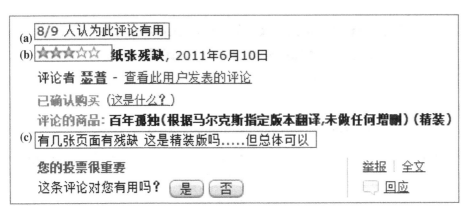

Fig. 1. Helpfulness of reviews from www.amazon.cn

three main factor combination methods with the information of helpfulness and conduct correlation analysis with sales rankings of books, so as to find the most correlative combinations, namely the optimal combinations. Helpfulness means assessing the quality of reviews by other users, namely judge whether the reviews are effective for the purchase decision [2-3]. An example of the books 'One Hundred Years of Solitude' is shown in Figure 1. There are eight of nine people who think the review is helpful; this book has 3 stars and the user thinks this book is generally nice, but may not be hardcover.

We aim at finding an effective combination for all domains and the specific combination for every domain, so as to help customers to find their favorite books. According to the experimental results, we can get the following conclusions: firstly, **WH (with helpfulness)** conclusion, which means that helpfulness is really useful to predict sales rankings of books; Secondly, **DOC (domain optimal combination)** conclusion, which shows that different domains of books have different domain optimal combinations; lastly, **OOC (overall optimal combination)** conclusion, which means that in the consideration of helpfulness, the combination of number of positive reviews, score of review stars and score of frequent aspects is the most correlative combination. All of these findings might be valuable information for customers to make the effective purchasing decisions.

The remain or rest of this paper is organized as follows. Section 2 reviews related works. Data collection and annotation are introduced in Section 3. Section 4 presents our methodology. Experimental results are provided in Section 5. The last part is about the conclusion and the future work.

2   Related Works

Two types of works are related to our study: sentiment analysis and sales forecast.

Sentiment analysis is to identify the attitudes of users by mining reviews. In this paper we focus on document-level and aspect-level sentiment analysis. Document-level sentiment analysis is to predict whether the whole document expresses a positive sentiment or a negative one [4]. Many researches have been done by supervised [5-7] and unsupervised learning methods [8-9]. Rather than gathering isolated opinions about a whole item, users generally prefer to compare specific features of different products, so it is important to conduct fine-grained aspect-level sentiment analysis [10]. Methods for sentiment analysis at this level are various. Methods like LDA models, sentiment lexicons are often used for aspect-level sentiment analysis [11-12]. In this paper, we use statistical methods to conduct document-level sentiment analysis, and lexical affinity methods for aspect-level sentiment analysis.

There are also many related works about sales forecast. Chang & Lai proposed a hybrid system to combine the self-organizing map of neural network with case-based reasoning method, for sales forecast of new released books [13]. Tanaka used high correlations between short-term and long-term accumulated sales within similar products groups to present a new forecasting model for new-released products [14]. Bollen's results indicated that the accuracy of DJIA (Dow Jones Industrial Average) predictions can be significantly improved by the inclusion of specific public mood dimensions [15]. Lee et al. developed and compared the performance of three sales forecasting models for the forecasting of fresh food sales, and the research results reveal that Logistic Regression performs better than the other methods [16]. Yu et al. conducted a case study in the movie domain to predict sales performance by analyzing the large volume of online reviews [17].

In this paper, we propose three main factor combination methods and conduct correlation analysis with book sales rankings, so as to find relationships between reviews polarities, helpfulness, stars and sales rankings of products.

3   Data

3.1   Data Collection

We collected sales rankings of three domains of books in the first half of 2013 from Amazon, including Literature[1], Social Science[2] and Economic Management books[3]. We chose top 50 books of each domain to conduct analysis. In total, we have collected 92,595 book reviews, including 60,903 literature book reviews, 17,476 social science book reviews and 14,216 economic management book reviews. The corpora cover reviews, stars and helpfulness of the books. The detail information is shown in Table 1.

---

[1] http://www.amazon.cn/gp/feature.html/ref=br_lf_m_353738_pglink_
1?ie=UTF8&docId=353738&plgroup=1&plpage=1

[2] http://www.amazon.cn/gp/feature.html/ref=br_lf_m_353748_pglink_
1?ie=UTF8&docId=353748&plgroup=1&plpage=1

[3] http://www.amazon.cn/gp/feature.html/ref=br_lf_m_353758_pglink_
1?ie=UTF8&docId=353758&plgroup=1&plpage=1

3.2   Data Annotation

In order to construct the training set, we tagged part of reviews manually. We have tagged 5, 000 reviews manually. Among them, 2, 500 reviews express a positive feeling towards the entity and 2, 500 reviews express a negative one. For the convenience and reliability of the further comparison, we conduct cross validation on the training set to test the performance. We employed SVM as the classifier. Specifically, we used the LibSVM[4] to conduct experiments with 5-fold cross validation and present evaluation results in Table 2. From Table 2 we can find that the performance of reviews annotation is excellent. Therefore, it is trustable to use it as training data to conduct sentiment analysis on the whole corpus.

Table 1. Samples of data collection

Domains	Books	Reviews	helpfulness	Stars
Literature	Insight	The whole book is like a novice worked out in a short period of time, bad writing, and unclear thinking.	45 / 48	1
Social Science	On China	Careful packaging. It is Content that matters. A good book.	3 / 3	5
Economic Management	Rich Dad, Poor Dad	Sorry. I know a lot of people like it, but I really don't love it.	0 / 1	2

Table 2. Cross-validation performance of the reviews annotation

Metrics	Recall	Precision	F1 value
Scores	0.9805	0.9756	0.9780

4   Methodology

4.1   Framework

We conducted correlation analysis of our combinations and sales rankings on book of three domains by combining the information of book reviews, review stars and review helpfulness. We proposed three main ranking schemes, each of them includes three parts: **without helpfulness**, which means that we would not take the information about helpfulness into consideration when we compute the book scores and sort them; **with helpfulness**, which means that the information about helpfulness would be taken into consideration; **product ranking**, we multiplied the book scores that we got from the above two steps and then sorted them. We conducted correlation analysis between our rankings and sales rankings, so as to find the optimal combinations. The details are shown in Figure 2.

---

[4] http://www.csie.ntu.edu.tw/~cjlin/libsvm/

4.2 Factor Combinations

In order to carry out the correlation analysis, we proposed 7 factor combination methods, which can be divided into three categories: combination 1, 2 and 3. The details are show in Table 3, and the calculations of factors are shown in Table 4.

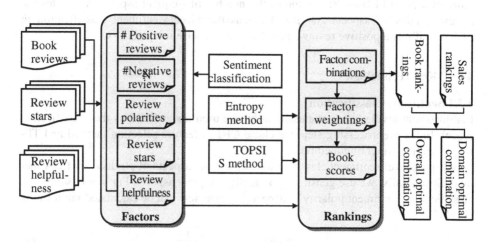

Fig. 2. Framework of optimal combinations selection

Table 3. Factor combination methods

Factors	Combination 1	Combination 2			Combination 3		
		2(a)	2(b)	2(c)	3(a)	3(b)	3(c)
#Positive reviews	O	O		O			O
#Negative reviews	O	O	O			O	
Score of review sentiment polarities	O		O	O	O		
Score of review stars	O	O	O	O	O	O	O
Score of frequent aspects	O	O	O	O	O	O	O

Table 4. Calculation of factors

Factors	Without helpfulness	With helpfulness
Score of review sentiment polarities	$scorep(B) = \sum_{i=1}^{N} sp(i)/N$	$scorep(B) = \sum_{i=1}^{N} sp(i)*he(i)/N$
Score of review stars	$scores(B) = \sum_{i=1}^{N} st(i)/N$	$scores(B) = \sum_{i=1}^{N} st(i)*he(i)/N$
Score of frequent aspects	$scorea(B)$ $= \sum_{i=1}^{m} (\sum_{i=1}^{N} po(i)/n)/m$	$scorea(B)$ $= \sum_{i=1}^{m} (\sum_{i=1}^{N} po(i)*he(i)/n)/m$

In Table 4, scorep(B) means score of review sentiment polarity of book B, $sp(i)$ means the sentiment polarity of review i, if it is a positive review, $sp(i)$ equals to +1, else it equals to -1; N denotes the number of reviews of book B, $he(i)$ donates score of helpfulness of review i. scores(B) donates score of review stars of book B; $st(i)$ means the star of review i, it ranges from 1 to 5. scorea(B) means score of aspects of book B; m denotes the number of frequent aspects, n denotes the review number of aspects A, $po(i)$ means the aspect sentiment classification in review i, if it is a positive review, $po(i)$ equal to 1, else it equal to -1.

4.3   Key Technologies

4.3.1 Sentiment Classification
For document-level sentiment classification, we used linear SVM as the classification model. In the preprocessing step, we chose CHI as feature selection method and TF-IDF as feature weighting method.

For aspect-level sentiment classification, we extract aspects of products by LDA method. Specifically, we use gensim [5] to identify frequent aspects. For aspect sentiment classification, sentiment polarity of aspect A in a review can be calculated via formula (1) [11].

$$Score(A) = \sum_{i=1}^{n} \frac{w_i SO}{dis(w_i, A)} \tag{1}$$

where w_i denotes a sentiment word, n means number of sentiment words in review s, and $dis(w_i, A)$ denotes the distance between aspect A and sentiment word w_i. $w_i SO$ is the sentiment score of the word w_i. If word w_i is a positive word, $w_i SO$ equals to +1, else it equals to -1. If $Score(A) > 0$, the sentiment polarity of aspect A in the review s is positive, else it is negative.

4.3.2 Factor Weighting Calculation
We use the entropy method to calculate factor weightings [18].
(1) Normalization
We calculate the proportion of object i in factor j, it is computed by Eq.(2)

$$P_{ij} = X_{ij} / \sum_{i=1}^{n} X_{ij}, (i = 1,2, \dots, n, j = 1,2, \dots, m \tag{2}$$

where, X_{ij} denotes value of object i in factor j; n means the numbers of books (here, it equal to 50, the same below); m means the numbers of factors.
(2) Factors entropies

$$e_j = -\frac{1}{\ln(n)} \sum_{i=1}^{n} P_{ij} \ln(P_{ij}) \tag{3}$$

where, e_j denotes entropy of factor j.
(3) Factor weightings

$$w_j = \frac{1 - e_j}{m - \sum_{j=1}^{m} e_j} \tag{4}$$

where, w_j denotes weighting of factor j; m means the numbers of factors.

---

[5] http://radimrehurek.com/gensim/

4.2.3 Book Score Calculation

We use the TOPSIS method to calculate book scores [19].

(1) Weighted factors

$$P_{ij} = w_j * P_{ij} \tag{5}$$

where, P_{ij} denotes value of weighted factors j of object i; P_{ij} means proportion of object i in factor j; w_j means weighting of factor j.

(2) Identification of ideal points

$$PIP_j = \max(P_{ij}) \, , (i = 1,2, ..., n) \tag{6}$$

$$NIP_j = \min(P_{ij}) \, , (i = 1,2, ..., n) \tag{7}$$

where, PIP_j denotes positive ideal point of factor j; n means the numbers of books (it equals to 50, the same below); NIP_j denotes negative ideal point of factor j.

(3) Distances of each book to the positive and negative ideal points

$$DP_i = \sqrt{\sum_{j=1}^{m}(P_{ij} - PIP_j)^2} \, , (i = 1,2, ..., n) \tag{8}$$

$$DN_i = \sqrt{\sum_{j=1}^{m}(P_{ij} - NIP_j)^2} \, , (i = 1,2, ..., n) \tag{9}$$

where, DP_i denotes distances of book i to the positive ideal points; NP_i denotes distances of book i to the negative ideal points; m means the numbers of factors.

(4) Score of each book

$$score(b_i) = NP_i/(DP_i + NP_i), \ (i = 1,2, ..., n) \tag{10}$$

where, $score(b_i)$ denotes score of book i.

5   Experiments

5.1   Overall Optimal Combination

5.1.1   Correlation Analysis on Combination 1

The results of correlation analysis on the combination 1 are shown in Table 5. From Table 5 we can find that, for Literature and Social Science, sales rankings and all the three rankings have significant correlations at the level of 0.01 (bilateral). Among them with helpfulness rankings have the biggest correlation coefficients. It means that this kind of ranking is more useful to predict sales rankings. However, for Economic Management, there is no significant correlation between sales ranking and our three rankings. All these analyses above show that combination 1 is not useful enough for all domains.

5.1.2   Correlation Analysis on Combination 2

The results of correlation analysis on combination 2(a) are shown in Table 6. The correlation results in Table 6 are similar to combination 1. So we can get the conclusion that combination 2(a) is not useful enough for all domains.

The results of correlation analysis on combination 2(b) are shown in Table 7. From Table 7 we can find that, for Literature and Social Science, sales rankings and all the three rankings have significant correlations at the level of 0.01 (bilateral). However, for Economic Management, there is no significant correlation between sales ranking and our three rankings. All these analyses above show that combination 2(b) is not useful enough for all domains.

The results of correlation analysis on combination 2(c) are shown in Table 8. From Table 8 we can find that, the correlation results are similar to combination

Table 5. Correlation analysis on combination one

Domains	Without Helpfulness	With Helpfulness	Product
Literature	0.348**	0.372**	0.360**
Social Science	0.381**	0.391**	0.389**
Economic Management	0.183	0.183	0.183

Table 6. Correlation analysis on combination 2(a)

Domains	Without Helpfulness	With Helpfulness	Product
Literature	0.303**	0.372**	0.365**
Social Science	0.372**	0.392**	0.382**
Economic Management	0.197	0.182	0.191

Table 7. Correlation analysis on combination 2(b)

Domains	Without Helpfulness	With Helpfulness	Product
Literature	0.332*	0.282*	0.335*
Social Science	0.361**	0.372**	0.369**
Economic Management	0.237	0.207	0.225

Table 8. Correlation analysis on combination 2(c)

Domains	Without Helpfulness	With Helpfulness	Product
Literature	0.405**	0.382**	0.389**
Social Science	0.384**	0.399**	.395**
Economic Management	0.241	0.223	0.224

2(b). So combination 2(c) is not useful enough for all domains.

5.1.3   Correlation Analysis on Combination 3
The results of correlation analysis on combination 3(a) are shown in Table 9. From Table 9 we can find that, for Literature and Social Science, sales rankings and last two rankings have significant correlations at the level of 0.01 (bilateral). However, for

Economic Management, there is no significant correlation between sales ranking and our three rankings. All these analyses above show that combination 3(a) is not useful enough for all domains.

The results of correlation analysis on combination 3(b) are shown in Table 10. From Table 10 we can find that, for Literature, only with helpfulness ranking and sales ranking have a significant correlation; For Social Science, sales ranking and all the three rankings have significant correlations at the level of 0.01 (bilateral), among them with helpfulness ranking have the biggest correlation coefficient. However, for Economic Management, there is no significant correlation between sales ranking and our three rankings. All these analyses above show that combination 3(b) is not useful enough for all domains.

The results of correlation analysis on combination 3(c) are shown in Table 11. From Table 11 we can find that, for Literature, sales ranking and all the three rankings have significant correlations at the level of 0.01 (bilateral); For Social

Table 9. Correlation analysis on combination 3(a)

Domains	Without Helpfulness	With Helpfulness	Product
Literature	0.256	**0.345****	**0.448****
Social Science	0.104	**0.384****	**0.328****
Economic Management	0.083	0.093	0.073

Table 10. correlation analysis on combination 3(b)

Domains	Without Helpfulness	With Helpfulness	Product
Literature	0.256	**0.372****	0.274
Social Science	**0.354****	**0.374****	**0.371****
Economic Management	0.236	-0.030	0.141

Table 11. Correlation analysis on combination 3(c)

Domains	Without Helpfulness	With Helpfulness	Product
Literature	**0.385****	**0.372****	**0.372****
Social Science	0.181	**0.401****	**0.372****
Economic Management	0.240	0.240	**0.368****

Science, sales ranking and last two rankings have significant correlations, and with helpfulness ranking have bigger correlation coefficient; For Economic Management, only product ranking and sales ranking have significant correlation. All the analysis above shows that product ranking in combination 3(c) is useful enough for all domains.

From the analysis above, we can draw the **OOC (overall optimal combination) conclusion** that combination 3(c) is the most useful combination, namely, in the consideration of helpfulness, the combination of numbers of positive reviews, score of review stars and frequent aspects is the most correlative combination.

5.2   Domain Optimal Combination

(1) Domain optimal combination of Literature books

We conducted correlation coefficients of three main combinations about Literature books and the results are shown in Figures 3. From Figure 3 we can find that the biggest correlation coefficient belongs to product ranking in combination 3(a), followed by combination 2(c) and 3(c). In addition, with helpfulness rankings are the highest of our proposed rankings in three of the combinations and product rankings are the highest in two of the combinations, which means that the information of helpfulness is useful to predict sales rankings of Literature books.

(2) Domain optimal combination of Social Science books

The results of correlation coefficients about Social Science books are shown in Figure 4. From Figure 4 we can find that the biggest correlation coefficient belongs to with helpfulness ranking in combination 3(c), followed by 2(c) and 2(a). With helpfulness rankings are the highest of our three rankings in all of the combinations, which means that helpfulness is useful to predict sales rankings of Social Science books.

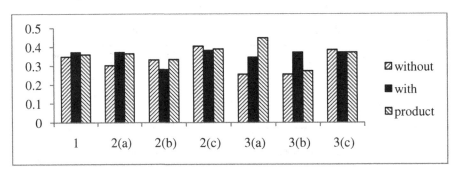

Fig. 3. Correlation coefficients of Literature books

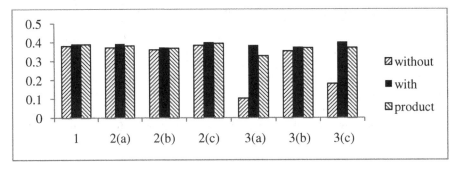

Fig. 4. Correlation coefficients of Social Science books

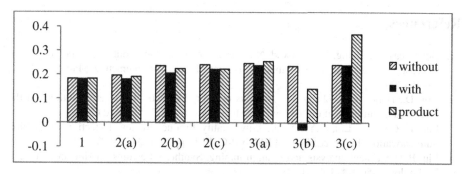

Fig. 5. Correlation coefficients of Social Science books

(3) Domain optimal combination of Economic Management books

For Economic Management, only product ranking in combination 3(c) has significant correlation with sales ranking, which also proved that helpfulness is useful to help predict sales rankings of Economic Management books.

From the analysis above, we can draw **DOC (domain optimal combination) conclusion** that different domains of books have different domain optimal combinations. For Literature books, the domain optimal combination is combination 3(a), while for Social Science and Economic Management books, combination 3(c) is the domain optimal combination. We can draw **WH (with helpfulness) conclusion** that helpfulness is really useful to predict sales rankings of books.

6 Conclusions and Future Works

In this paper, we proposed three main factor combination methods and chose the optimal ones via correlation analyses between our combinations and book sales rankings. Three main conclusions can be drawn according to our above mentioned analysis:

(1) **WH conclusion**: the information of helpfulness is really useful to predict or help predict sales rankings of books.
(2) **DOC conclusion**: different domains of books have different domain optimal combinations.
(3) **OOC conclusion**: in the consideration of helpfulness, the combination of numbers of positive reviews, score of review stars and score of frequent aspects is the most correlative combination.

The data in this paper is in Chinese, however our method for classification and correlation analysis is language independent. According to the three conclusions, we may predict sales rankings of books and provide effective purchasing suggestions for customers. In the future works, we will consider more languages of book and more types of products in the future. In addition, we will filter the untrusted reviews more efficiently.

Acknowledgments. This work is supported by Major Projects of National Social Science Fund (13&ZD174), National Social Science Fund Project (No.14BTQ033) and the Opening Foundation of Alibaba Research Center for Complex Sciences, Hangzhou Normal University (No. PD12001003002003).

References

1. Dellarocas, C., Zhang, X.M., Awad, N.F.: Exploring the value of online product reviews in forecasting sales: The case of motion pictures. Journal of Interactive Marketing 21(4), 23–45 (2007)
2. Yin, D., Bond, S.D., Zhang, H.: Anxious or angry? Effects of discrete emotions on the perceived helpfulness of online reviews. Mis Quarterly 38(2), 539–560 (2014)
3. Liu, J., Cao, Y., Lin, C.Y., et al.: Low-Quality Product Review Detection in Opinion Summarization. In: Proceedings of EMNLP-CoNLL, pp. 334–342 (2007)
4. Liu, B.: Sentiment analysis and opinion mining. Synthesis Lectures on Human Language Technologies 5(1), 1–167 (2012)
5. Mullen, T., Collier, N.: Sentiment Analysis using Support Vector Machines with Diverse Information Sources. In: Proceedings of EMNLP, pp. 412–418 (2004)
6. Xia, R., Zong, C., Li, S.: Ensemble of feature sets and classification algorithms for sentiment classification. Information Sciences 181(6), 1138–1152 (2011)
7. Li, S., Lee, S.Y.M., Chen, Y., Huang, C.-R., Zhou, G.: Sentiment classification and polarity shifting. In: Proceedings of the 23rd International Conference on Computational Linguistics, pp. 635–643 (2010)
8. Turney, P.D.: Thumbs up or thumbs down?: Semantic orientation applied to unsupervised classification of reviews. In: Proceedings of the 40th Annual Meeting on Association for Computational Linguistics, pp. 417–424 (2002)
9. Taboada, M., Brooke, J., Tofiloski, M., Voll, K., Stede, M.: Lexicon-based methods for sentiment analysis. Computational Linguistics 37(2), 267–307 (2011)
10. Cambria, E., Hussain, A., Havasi, C., Eckl, C., Munro, J.: Towards crowd validation of the UK national health service. In: Proceedings of WebSci, pp. 1–5 (2010)
11. Ding, X., Liu, B., Yu, P.S.: A holistic lexicon-based approach to opinion mining. In: Proceedings of the International Conference on Web Search and Data Mining, pp. 231–240 (2008)
12. Moghaddam, S., Ester, M.: On the design of LDA models for aspect-based opinion mining. In: Proceedings of the 21st ACM International Conference on Information and Knowledge Management, pp. 803–812 (2012)
13. Chang, P.C., Lai, C.Y.: A hybrid system combining self-organizing maps with case-based reasoning in wholesaler's new-release book forecasting. Expert Systems with Applications 29(1), 183–192 (2005)
14. Tanaka, K.: A sales forecasting model for new-released and nonlinear sales trend products. Expert Systems with Applications 37(11), 7387–7393 (2010)
15. Bollen, J., Mao, H., Zeng, X.: Twitter mood predicts the stock market. Journal of Computational Science 2(1), 1–8 (2011)
16. Lee, W.I., Chen, C.W., Chen, K.H., et al.: A comparative study on the forecast of fresh food sales using logistic regression, moving average and BPNN methods. Journal of Marine Science and Technology 20(2), 142–152 (2012)
17. Yu, X., Liu, Y., Huang, X., et al.: Mining online reviews for predicting sales performance: A case study in the movie domain. IEEE Transactions on Knowledge and Data Engineering 24(4), 720–734 (2012)
18. Hongzhan, N., Lü Pan, Q.Y., Yao, X.: Comprehensive fuzzy evaluation for transmission network planning scheme based on entropy weight method. Power System Technology 33(11), 60–64 (2009)
19. Hwang, C.L., Yoon, K.: Multiple Attribute Decision Making: Methods and Applications, A State of the Art Survey. Springer, New York (1981)

Research on Webpage Similarity Computing Technology Based on Visual Blocks

Yuliang Wei, Bailing Wang[*], Yang Liu, and Fang Lv

Harbin Institute of Technology, Weihai, Shandong, 264209
wbl@hit.edu.cn

Abstract. Measuring web page similarity is one of the core issues in web content detection and Classification. In this paper, we first give the definition of webpage visual blocks. And then we propose a method using visual blocks for measuring web page similarity. The experiments show our method can effectively measure level of similarity between different type of webpages.

Keywords: Visual Blocks, Visual Comparison, Structure Comparison, Web Page Similarity.

1   Introduction

With the rapid development of the Internet and the growing network of multimedia data, social networking and other forms of networking applications are common and famous. More and more daily lives are mapped to network information. Mining the laws of the social life from the multifarious and numerous network information is a hot area of sociology of science. The webpage is an important way to show information. Webpage information processing has a fundamental role in social network information processing. Analysis of webpage structure is one of great significance area in webpage information processing. This paper researches the relationship between the similarity of pages on the basis of similar visual blocks of the page.

Webpage similarity analysis is widely used in information processing of social media, like duplicate removal of collection results, phishing detection (see [1]), webpage tamper detection, automatic information extraction(see [2]) and so on. Traditional webpage similarity analysis can be divided into text similarity calculation and structural similarity calculation.

Text similarity calculation establishes computable representation model based on text messages through extracting the main text message of the content. Text similar calculation is used to solve webpage text similarity problems (see detail in [3]).

Structural similarity calculation establishes computable representation model based on webpage structural by analyzing the structural properties of the webpage visual to solve structural similarity problems.

This paper combines the two methods above-mentioned and proposes a similarity calculation method based on visual blocks. This method can be used both for phishing

---

[*] Corresponding author.

H.-Y. Huang et al. (Eds.): SMP 2014, CCIS 489, pp. 187–197, 2014.

detecting, mirror site detection, web tamper detection and the relation between picture, text and other elements in a webpage.

2   Related Work

Webpage similarity calculation based on visual structure has a basic role in many web-based computing applications. The similarity measure has slightly different on same pages in different occasions. Phishing detection, web content automatically extracting and many other web computing problems use web similarity for solving in previous works.

In paper [4], Zhang's group calculate the similarity between legitimate web page and phishing web page. To do this job, they get the visual area spatial by analyzing DOM-Tree and get layout features using spatial layout feature extraction based on image segmentation. Then they calculate the similarity using vision-block and overlap area ratio of two blocks. According to results of imulation experiments, this method performances well in phishing detection, but it compares every blocks in the webpage, making it not adapted to dealing with mass data.

Yasufumi's groups proposed method compares the layouts of Web pages based on image processing and graph matching. The experimental results show that the accuracy of layout analysis is 91.6% in average (see [5]). While it is very time-consuming when transform a web page to image and the method don't consider the textural feature.

Law's group learn the similarity between Web pages using image and structural techniques (see [6]). They get images in webpage and visual structures by analyzing DOM-Tree and rendering webpage, and classify webpages into different kinds of classifications, whose accuracy is up to 93.1%, using VI-DIFF algorithm.

Besides, Law's group analyze the change with time of one webpage, based on visual features and structural techniques. They get visual blocks using image of webpage's rendering capture and comparing the visual features in each page version, which proves that two versions of a webpage are considered similar if the changes that occurred between them are not important enough to archive both of them, only using the visible part of web pages without scrolling the vertical scroll bar. This page calculated similarity of webpages only based on image using complex OCR algorithm with high computational complexity, and did not give the numerical similarity of webpages.

Marinai's group also used visual structure for web page classification (see in [7]). They found that in most cases people didn't need to read the contents, but can know what kind of category did the articles belong to. Such as news, business documents, papers or other type. They present an alignment method of page visual block tree matched to achieve similarity calculated by the page layout, and for comparison. Their method is suitable for one to one comparison. The same methods can be found in [8].

In paper [9], researchers propose a similarity calculation which use both the page content and visual blocks. They discuss separately about text similarity and visual structure similarity.

Studies before all used concept of block, but not giving the definition of block. In this page we make a definition of block at first, and give an algorithm to select the best visual-block. Also, we give an algorithm to calculate the similarity between visual-blocks, considering both visual-block's structure feature and content feature, performancing more comprehensively compared to algorithms only based on structure feature or image. Furthermore, we use location as key value when matching blocks in this page, improving the efficiency of match a lot.

3   Webpage Similarity Calculation Based on Visual Blocks

We analyzed the relationship between the DOM tree nodes and page layout and gave a definition of the page visual blocks with a filter algorithm for selecting node from DOM tree. In order to improve the accuracy of similarity, our method compared with previous researches takes into account the page visual structure and the content features of visual blocks. In addition we found that there is two different types of visual block for a page, herein referred to them as static and dynamic visual block. Static visual block refers to blocks which content does not change within a certain time visual. The visual dynamic blocks are those blocks will be changed.

In Fig. 1, the area which tag one is belong to navigation area for the entire page. For a page navigation area will not change in a long time which is relatively stable. So the area one is a static visual block. As for the tag two area, it is a news links area. News will be dynamically updated over time. The content is constantly changing. So it is belong to dynamic block.

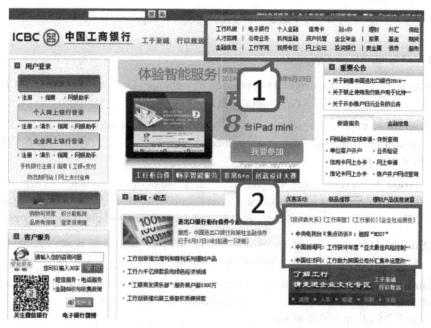

Fig. 1. Example of Vision Block

According to the definition and selection of visual block algorithm we proposed, Fig. 1 will be resolved to Fig. 2. The similarity calculation we proposed is rely on visual similarity comparison of the divided blocks.

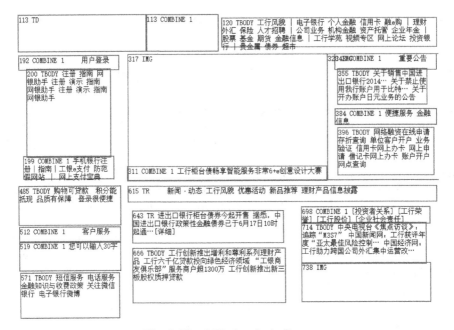

Fig. 2. Visual Blocks of the Page

3.1   Visual Block Extraction

Set up W represent any original information of webpage. Dt represents the DOM tree structure of the web W. Nt represent nodes in a tree structure Dt collection.

Definition 1: ∀dti ∈ Dt, set up pi = <hi, vi> is the location information of dti. Then the visual tree of the web W can be defined as Vt = {v0, v1, v2 …}, in which ∀vi∈ Vt, vi = <dti, pi>.

Definition 2: For visual tree Vt={n0, n1, n2 …}, we defines f(ni) = nj　（∀ni，nj ∈ Vt）as nj is the direct parents of ni. For collection Nc, if ∀ni∈Nc, f(ni) = nj and ∄n ∈ Vt，n ∉ Nc, meet f(ni) = nj, we call Nc is complete set of child nodes for nj.

Definition 3: For the set of nodes Nv, if ∀ ni,ni∈Nc，f(ni) = nj is not established, Nv is called as a visual block partition of a web page.

Definition 4: Set up Vt's node ni, a visual block division Nv ' and ni's all direct child nodes collection Nci'. We give a cover operation' definition on Vt.

if ni∈Nv', then Nv' is ni's cover.

if ∀nj∈Nc', Nv'v is nj's cover, then ni is covered by Nv'.

Deduction 1: Set up function F(Nv)=N（Nv，N⊂Vt）which represents the nodes collections covered by Nv is N. So for a node ni, if Nci⊂N，then ni∈N.

Definition 5: Set up F(Nv)=N（Nv，N⊂Vt），if the root node n0∈N, we call Nv is a visual block collection of the page marked symbol Vb.

Nature 1: For Vt and Vb of one page, all paths starting from the root of Vt to the leaf nodes are cut by nodes in Vb.

Vb has the nature 1 shows that nodes in Vb can cover the entire visible area of the page. Vb for one page is not only one. We need to find a suitable Vb collection so it can represent the hole page. Let conditional filtering function C(X, n) →(0,1), X is a vector of filters.

Definition 6: Optimal Vb collection of page.

∀ni∈Vb，if C(X,ni)=1 then ∀nj∈Nci，C(X,nj)=0,

∀ni∈Vb，if C(X,ni)=0 then f(ni)=nif, Ncif 属于 Vb,

Actually in this paper, the visual block extraction is to find the optimal Vb of page. The visual block filters including:

1. Node area greater than a threshold value.
2. Merging similar or related area, such as continuous text, links, blocks, etc.
3. Visual content between the adjacent blocks are relatively independent.

There we give the filter algorithm description. The algorithm uses Recursive traversal to find nodes which meet the qualifications.

Algorithm input：	Vt of web page
Algorithm output：	Vb

```
1:    def check_node(node):
2:        if node['name'] in IGNORE_LIST:
3:            return False
4:        if node['childs']:
5:            if len(node['text']) > sum(len(node['childs']['text'])):
6:                select_nodes.append(node)
7:                return True
8:            for cnode in node['childs']:
9:                check_node(cnode)
10:           if len(select_cnodes) == len(node['childs']):
11:               node['select'] = 1
12:               return True
13:           elif len(select_cnodes) > 0:
14:               slice_childs = slice(node_childs by select_cnodes)
15:               for slice_list in slice_childs:
16:                   combine_node = combine(slice_list)
17:                   if combine_node:
18:                       select_nodes.append(combine_node)
19:                   else:
20:                       return judge(node)
21:           else:
22:                   return judge(node)
```

3.2   Visual Block Feature Extraction and Visual Block Matching Calculation

We can get an optimal Vb from page using the algorithm in 3.1. To achieve comparison between pages, we needs to extract content features from each visual block of Vb. In this paper, the visual content features of the selected blocks include:

1. The node position information, np(top, left, height, width). The location information includes the width and height of the area occupied by the node and relative offset of the upper left corner of node to the upper left corner of the page.
2. Text content of the node (including the text of the child tree nodes), text.
3. The ratio of the number of A label node in the child tree nodes and the node height, lp, see Equation (1). The ratio is a measure of the visual block to link density. 20 in formula is the default text height in pixels of webpage.

$$lp = \frac{linkNum \times 20}{heigh} \qquad (1)$$

4. The ratio of image area in node area and the node area, pp, Equation (2).

$$pp = \frac{pictureArea}{heigh \times width} \qquad (2)$$

5. The number of input label in child tree nodes, ins.

A visual area can be described with these features. Similar calculation of visual block mainly dependents on these features. In addition to these features of visual block, we need to identify the static blocks in Vb.

We can use whether np is equality for judging position information matching of two visual blocks. Actual we only use height, width and left of the visual block, because the value of top is not stable for a page. If the top is used to compare, the same blocks will not match. For the text of visual block, we use Jaccard Similarity(see in [10]) method to calculate the similarity. For the other three features, we introduce the variable X for similarity calculation, show in Equation (3).

$$x = 1 - \frac{2v_i}{v_i + v_j} \qquad (3)$$

Where vi and vj are the feature values of the blocks to be detected and vj is the matched page. Then the feature matching value y can be used by normal distribution function, see Equation (4).

$$y = \frac{1}{\sigma\sqrt{2\pi}} e^{\frac{-(x-\mu)^2}{2\sigma^2}} \qquad (4)$$

Where $\sigma = 1, \mu = 0$.

The calculation of visual matching M between the visual block vi which to be matched and the default visual block vj is defined as follows, see Equation (5).

$$M(v_i, v_j) = \frac{\sum y_i}{5}$$ (5)

A page will be generally divided into 50 to 150 pieces of visual blocks, depending on the size of the page. On phishing detecting, there are over thousands of pages need to be protected. We need to process the pages to visual blocks before comparison. So we establish the inverted index for inquiring. Considering the relationship between the index and contents, we use the position information as the index key, the others as contents. The index relationship is shown in Fig. 3.

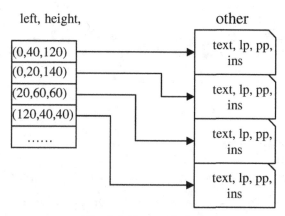

Fig. 3. Indicate of Index

3.3   Visual Block Matching and Page Similarity Calculation

The visual similarity of pages depend on the similarity of each visual block from page. Both static blocks and dynamic blocks are considered. For page structure block comparison, the static and dynamic blocks have the same weights. While in phishing detection the static blocks are more important than dynamic blocks. If we consider removing duplicate by page content or mirror site detection, we should increase the weight of dynamic blocks. Based on the above analysis, set default page P1 and the visual blocks of P1 is Vb1. The page to be detected is P2, and the visual blocks of P2 is Vb2. The similarity is defined as Equation (6).

$$S(P_1, P_2) = \frac{\alpha S_s(VB_1, VB_2) + \beta S_o(VB_1, VB_2)}{len(VB_1) + len(VB_2)} \times 2$$

$$= \frac{\alpha \sum_{VB_{1s}, VB_{2s}} M(vb_i, vb_j) + \beta \sum_{VB_{1o}, VB_{2o}} M(vb_i, vb_j)}{len(VB_1) + len(VB_2)} \times 2$$ (6)

Where the position value(left, width, height) of vbi and vbj are equal. Ss is function for calculating the match degree of static block, while So is the dynamic match degree. α and β are the weights of static and dynamic visual blocks.

4   Experiment

We use 20 consecutive days page information of popular portals. Then we set the first day page as default. The other days information are comparative data which is used to calculate the visual similarity with the first day. The   shows the result.

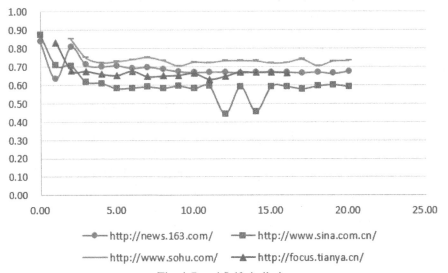

Fig. 4. Portal Self-similarity

There are two points worth noting in Figure. First, on the first day when x is 0, default page compared to itself, the visual similarity is 0.9, not 1. This is because when we calculate the visual similarity of the blocks, we don't know which block is the exact one of the default page. All of the blocks are indexed by position value, we can only use the triad (left, height, width) to shrink the range lookup. We just find the best matched block by order, so we can't make the similarity 1, but close. The other worth noting point is the similarity of the page compared to the first day page is reduced over time. This is because the content of the page is changing as times goes on. The greater change is with the longer time. So the similarity is reduced over time. There is a limit on reducing. The similarity is greater than 0.5. Because the layout of a page is stable, the structure will not change very soon.

In part 3.2, we don't use the top in position value in similarity calculation, because its value is changing for one visual block. The length of content in visual block or some repeat structure, like the link number of the area is not fixed, is change. So the top value of relevant visual blocks are change. The Fig. 5 show when we don't remove the top value, the similarity curve.

When we compute the similarity of the two pages, we use the position value as index. If we use the top value, the matching result is confusion and similarity is wrong.

In Fig. 6, we chose two different pages and calculate their degree of similarity. The result show the similarity is lower to zero, because different pages has different layout. This can prove our method is effective on similarity calculation of pages.

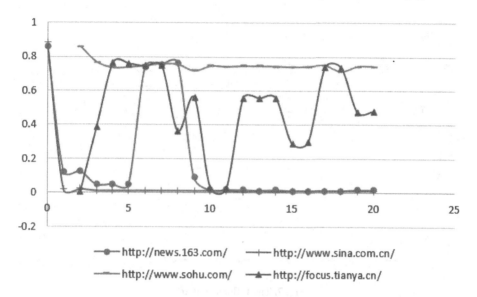

Fig. 5. Similarity Include Top

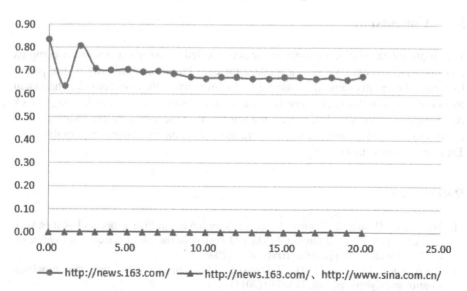

Fig. 6. Similarity Calculation of Different Pages

At the end, we choose different weight of static and dynamic visual blocks, α and β. We increases the value ofαIn turn, whileβ is equal to 1-α. The result shows in Fig. 7. With the increase of the value of α, the similarity of the pages is greater. Because our experiments are test on the same page. On other occasions, increase of the value of α may cause the result lower.

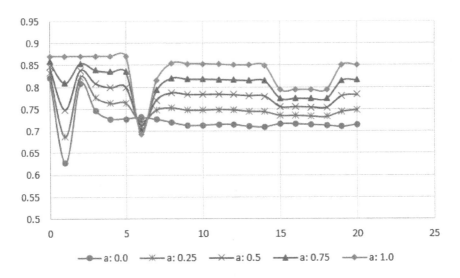

Fig. 7. The Influence of α

5   Conclusion

The similarity calculation of pages are widely applied in today's increasingly complex network environment. We give a method for calculating the similarity of the pages. Experimental results show that our method is effective. We only consider the visual blocks of pages in this paper. The types of visual blocks, like text block, image block etc. are missed. We also don't select the important in the visual block, which can use in advertising recognition, content extraction, automated extraction template etc. These are our next tasks.

References

1. Wenyin, L., Huang, G., Xiaoyue, L., et al.: Detection of phishing webpages based on visual similarity. In: Special Interest Tracks and Posters of the 14th International Conference on World Wide Web, pp. 1060–1061. ACM (2005)
2. Baczkiewicz, M., Łuczak, D., Zakrzewicz, M.: Similarity-based web clip matching. Control and Cybernetics 40, 715–730 (2011)
3. Salton, G., Buckley, C.: Term weighting approaches in automatic text retrieval. Information Processing and Management 24, 513–523 (1998)

4. Zhang, W., Lu, H., Xu, B., et al.: Web phishing detection based on page spatial layout similarity. Informatica 37(3), 231–244 (2013)
5. Takama, Y., Mitsuhashi, N.: Visual similarity comparison for Web page retrieval. In: Proceedings of the 2005 IEEE/WIC/ACM International Conference on Web Intelligence, pp. 301–304. IEEE (2005)
6. Law, M.T., Gutierrez, C.S., Thome, N., et al.: Structural and visual similarity learning for Web page archiving. In: 2012 10th International Workshop on Content-Based Multimedia Indexing (CBMI), pp. 1–6. IEEE (2012)
7. Marinai, S.: Page Similarity and Classification. In: Handbook of Document Image Processing and Recognition, pp. 223–253 (2014)
8. Cai, D., Yu, S., Wen, J.-R., Ma, W.-Y.: Block-based Web Search. In: The 27th Annual International ACM SIGIR Conference on Information Retrieval, pp. 440–447. ACM, Sheffield (2004)
9. Bartík, V.: Measuring web page similarity based on textual and visual properties. In: Rutkowski, L., Korytkowski, M., Scherer, R., Tadeusiewicz, R., Zadeh, L.A., Zurada, J.M. (eds.) ICAISC 2012, Part II. LNCS, vol. 7268, pp. 13–21. Springer, Heidelberg (2012)
10. Thada, M.V., Joshi, M.S.: A Genetic Algorithm Approach for Improving the average Relevancy of Retrieved Documents Using Jaccard Similarity Coefficient. International Journal of Research in IT & Management 1(4) (2011)

Doctor Recommendation via Random Walk with Restart in Mobile Medical Social Networks

Jibing Gong[1,2], Ce Pang[3], Lili Wang[2], Lin Zhang[2], Wenbo Huang[2], and Shengtao Sun[1,2]

[1] School of Information Science and Engineering, Yanshan University, Qinhuangdao 066004, China
[2] The Key Laboratory for Computer Virtual Technology and System Integration of Hebei Province, Yanshan University, Qinhuangdao 066004, China
gongjibing@163.com
[3] Department of Electronic Information and Science, College of Science, Yanshan University, Qinhuangdao 066004, China

Abstract. In this paper, we try to systematically study how to perform doctor recommendation in mobile Medical Social Networks (m-MSNs). Specifically, employing a real-world medical dataset as the source in our study, we first mine doctor-patient ties/relationships via Time-constraint Probability Factor Graph model (TPFG), and then define the transition probability matrix between neighbor nodes. Finally, we propose a doctor recommendation model via Random Walk with Restart (RWR), namely RWR-Model. Our real experiments validate the effectiveness of the proposed method. Experimental results show that we obtain the good accuracies of mining doctor-patient relationships from the network, the performance of doctor recommendation is also better than the baseline algorithms: traditional Reduced SVM (RSVM) method and IDR-Model.

Keywords: Doctor Recommendation, Random Walk with Restart, Doctor-patient Ties Mining, Mobile Medical Social Networks.

1 Introduction

Mobile Medical Social Networks (m-MSNs) play an increasingly important role in people's healthcare. How to mine and analyze m-MSNs is a hot research issue and recently attracts much attention of both industry and research communities. There have been a few studies on social recommendations. However, they almost completely ignored insufficiency of real medical information, the heterogeneity and diversity of the social relationship. In this paper, we try to study the problem how to perform doctor recommendation in a real-world m-MSN. Specifically, we first make doctor-patient relationship mining via Time-constraint Probability Factor Graph model (shortly TPFG)[1], and then define the transition probability matrix between neighbor nodes. Finally, we propose a doctor recommendation model via Random Walk with Restart (RWR) in a real-world m-MSN[2]. Experimental results show that

H.-Y. Huang et al. (Eds.): SMP 2014, CCIS 489, pp. 198–205, 2014.

we obtain good accuracy of mining doctor-patient relationships from the network, the doctor recommendation model is also better than baseline methods.

In this paper, we try to systematically investigate the problem of doctor recommendation in a real-world m-MSN with the following contributions:

- We make mining and analysis, especially the problem of doctor recommendation, in a real-world mobile Medical Social Network (m-MSN).
- We mine doctor-patient relationships via Time-constraint Probability Factor Graph mode (TPFG) from a m-MSN.
- We propose a novel model via Random Walk with Restart (RWR), namely RWR-Model, for doctor recommendation.

2   Our method

2.1   Doctor-Patient Ties/Relationships Mining via TPFG

Doctor-patient ties/relationships mining is the basis of accurately recommending doctors. The task can be formalized as: input time-correlated cooperation relation network $G = \{(V = V^p \cup V^a, E)\}$, where $V^p = \{p_1, ..., p_{n_p}\}$ denotes the set of disease case, the curing time of p_i is expressed by t_i, $V^a = \{a_1, ..., a_{n_a}\}$ stands for the set of all participants during curing disease, as well as E is the edge set each $e_{ji} \in E$ of whose edges links p_i and a_j together, and e_{ji} denotes a_j is one of all participants in the disease case p_i. The output of this model is a directed acyclic graph $H = (V', E'_s, \{(r_{ij}, st_{ij}, ed_{ij})\}_{(i,j) \in E'_s})$, where H is a subgraph of G and $E'_s \in E'$. In the edge-correlation information $(r_{ij}, st_{ij}, ed_{ij})$, r_{ij} indicates the probability that the participant a_j is a doctor a_i, as well as st_{ij} and ed_{ij} denote the starting time and end time of doctor-patient relationship duration, respectively.

This paper applies Time-constraint Probability Factor Graph model (shortly TPFG) [6] to mine doctor-patient ties/relationships. In this model, for each node a_i, three variables y_i, st_i, ed_i need to be solved. Given region feature function $g(y_i, st_i, ed_i)$, to reflect all joint probabilities in the relational graph, we define the joint probability as the product of all region feature functions, as in (1):

$$P(\{y_i, st_i, ed_i\}_{a_i \in V^a}) = \frac{1}{Z} \prod_{a_i \in V^a} g(y_i, st_i, ed_i) \tag{1}$$

where $1/Z$ indicates the normalization factor of the joint probability, and there two basic assumptions exist: (1) $\forall 1 \le x \le n_a, py_{yx}^1 < py_x^1$ indicates that a patient knows fewer about the disease than his/her candidate doctor. (2) $\forall a_i \in V^a, ed_{yi} < st_i < ed_i$ tells that a patient obtains diagnosis result/information later than his/her doctor. In this equation, to obtain most probable values of all unknown factors, the joint probability need to be maximized. A great deal of unknown parameters would lead to too large-scale solution space. Therefore, we simplify the joint probability problem into the following equation: Suppose if the patient a_i and his/her doctor y_i are determined, so we can obtain $\{st_i, ed_i\} = \arg\max_{st_i < ed_i} g(y_i, st_i, ed_i)$, that is, st_i and ed_i can be well solved. Before working out this joint probability, we first compute st_i and ed_i which are contained in every possible pair of doctor-patient relationship, and then we can get joint probability formula with simplified parameters, as in (2).

$$P(y_1, y_2 \cdots y_{n_a}) = \frac{1}{z} \prod_{i=1}^{n_a} f_i(y_i | \{y_x | x \in Y_i^{-1}\})$$

$$f_i(y_i | \{y_x | x \in Y_i^{-1}\}) = g(y_i, st_{ij}, ed_{ij}) \prod_{x \in Y_i^{-1}} I(y_x \neq i \vee ed_{ij} < st_{xi}) \qquad (2)$$

$$I(y_x \neq i \vee ed_{ij} < st_{xi}) = \begin{cases} 1, y_x \neq i \vee ed_{ij} < st_{xi} \\ 0, y_x = i \wedge ed_{ij} \geq st_{xi} \end{cases}$$

After simplifying formula (1), we can use probability factor graph[7] to solve formula (2). The graph includes two types of nodes: variable nodes and function nodes. Variable nodes y_i are corresponding to hidden variables $\{y_i\}_{i=0}^{n_a}$. Every variable node links one function node $f_i(y_i | \{y_x | x \in Y_i^{-1}\})$, which indicates $f_i(y_i | \{y_x | x \in Y_i^{-1}\})$ is determined by y_i. To reduce the time and space cost performed on the TPFG-based method, we design the rules and corresponding algorithm to filter those connections that do not stand for doctor-patient cooperation relationships.

2.2   Transition Matrix Definition

Given every participant node u, we need to obtain its neighbors to build a social network. Given $<u, v>$ of every directed edge, we define transition probability from u to v as in (3).

$$p(u,v) = log(\#uv+1.1)*log(\#vu+1.1) \qquad (3)$$

where $\#uv$ denotes the number of doctor u's checking patient v during the course of diagnosis and treatment, $\#vu$ states the number of patient v visiting to doctor u.

We can obtain the Intimacy Transition Probability Matrix (ITP-Matrix) according to Eq. (3), and use ITP-Matrix as the probability matrix of u's random-walker with restart. Considering the over-convergence problem in Eq. (3), we further improve it by introducing the divergence factor $C\ (u,\ v)$ as in (4). Then the new transition probability can be computed by the Eq. (5).

$$C\ (u,\ v) = |F\ (u)\ \cap\ F\ (v)|\ /\ C\_MAX \tag{4}$$

$$p_{new}\ (u,\ v) = \lambda p\ (u,\ v) +\ \gamma \cdot C(u,v) \tag{5}$$

where $F(u)$ and $F(v)$ denotes the node set of u's and v's neighbors respectively, $C\_MAX(=2000)$ is a constant for standardization, λ and γ are weights, and their values are 0.5.

2.3   Doctor Recommendation via Random Walk with Restart

To rank the nodes, we perform random walk with restart on a m-MSN at node u and calculate Ranking Score i according to Eq.(6)

$$RW(u_i) = \alpha \cdot RW(u_i) + \beta \cdot \sum_{u_j \in R_{u_i}} p(u_i, u_j) * RW(u_j) \tag{6}$$

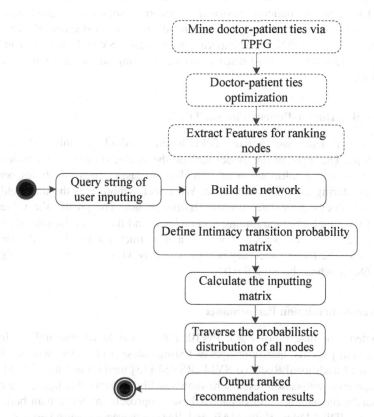

Fig. 1. The algorithm flow diagram of doctor recommendation

where $RW(u_i)$ denotes the set of all neighbor nodes of u_i in the network. $\alpha(=0.25)$ and $\beta(=0.75)$ are weighted values. Fig. 1. shows the algorithm flow diagram of our methods. After performing random walk, every node will have a ranking value which is regarded as the recommendation score of every node.

3   Experiments and Evaluations

3.1   Data Set

We have developed PDhms[2], a WSNs-based and mobile healthcare monitoring system for human pulse diagnosis. This system has helped us to collect the real-world medical dataset with 2064 size in a large scale clinic experiments which happened at the Institute of Computing Technology (ICT), Chinese Academy of Sciences (CAS) in 2009 and 2010, as well as the Hitech Fair of China in Shenzhen City in 2010 and 2011, respectively. It involves many kinds of diseases, doctor information, curing or treatment information, patients' personal information, colleague relationship, doctor-patient relationship and patients' evaluation information on doctors. Hence, we can build a real-world m-MSN. In the experiments, to train RSVM, IDR-Model and RWR-Model, we select 1064 of 2064 disease cases as training dataset and the rest cases as testing dataset.

3.2   Mining Doctor-Patient Ties via TPFG

In this experiment, we utilize TPFG-based method to mine doctor-patient relationships. The input of the model include the twelve doctor's information in our medical dataset, the information of 256 disease cases, diagnosis times, symptom information during treatment and so on. We select symbol θ as the threshold telling whether one doctor-patient relationship is true or not. The greater the value of the threshold θ is, the higher the mining accuracies is and the lower the rate of recalls is. In the experiment, we select $\theta = 0.7$ and extract totally 1210 doctor-patient relationships. The mining accuracy is approximately 81.3%. It is relatively high since it is just 68.5% when choosing $\theta = 0.6$.

3.3   Recommendation Performance

We adopted four indexes of P@5, P@10, P@15, and MAP, and utilizes feedback information of patients' questionnaires as testing dataset to compare with the baseline algorithms of traditional Reduced SVM (RSVM)[15] method and the IDR-Model[3]. The comparative evaluation results are shown in Fig. 2. From this figure, we can find recommendation accuracies of our RWR-based approach are better than both RSVM method and IDR-Model. Both MAE and RMSE metrics are used to measure the recommendation quality. From Table 1, the proposed RWR-based approach consistently outperforms the other two methods.

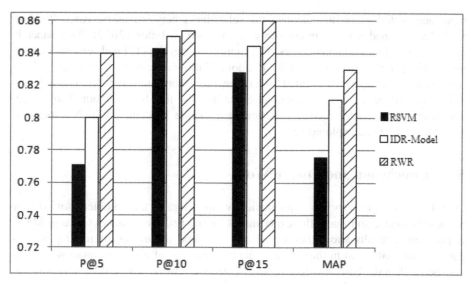

Fig. 2. Comparative evaluation results

Table 1. Performance comparison on MAE and RMSE in a real-world medical dataset

Method	MAE	RMSE
RSVM	0.28534	0.32865
IDR-Model	0.27425 ±(0.00231)	0.31264 ±(0.00300)
RWR	**0.22543±(0.00367)**	**0.21487 ±(0.00312)**

4   Related Work

In a Medical Social Network (MSN), but not a *m-MSN*, Ref.[3] proposed IDR-Model, an individual doctor recommendation model via weighted average method. Both application background and operating principle of the model are distinguished from this paper. Also, a closely related research topic is expertise search, such as expertise search based on candidate vote by Macdonald and Ounis[4], expertise mining from social network by Tang et al.[5], and transfer learning from expertise search to Bole search by Yang et al.[6]. In the literature, some recommendation methods only consider the similarity between the query keywords, but ignore the relationships between doctors and patients in a social network. Other relevant methods include low-rank matrix factorization models[7], content-based method[8], collaborative filtering[9], and Model-based approach[7]. To handle very large data sets, Ref.[10] presents a class of two-layer undirected graphical models, called Restricted Boltzmann Machines (RBM's). Ref.[9] develop a joint personal and social latent factor (PSLF) model for social recommendation.

Random Walk plays an important role in many fields and has attracted a lot of research interest. Jie Tang et al.[11] performed   a Random Walk with restart on the topic augmented graph to calculate the relatedness between users. Also, Random Walk have gained a lot of interest in academic search fields[12][13]. This paper is mainly inspired by two recent work on graph-based learning[12] and semi-supervised learning[14]. Ref.[14] performed the Random Walk with restart for personalized tag recommendation and incorporated Heterogeneous Information for Personalized Tag Recommendation in Social Tagging Systems. To the best of our knowledge, researches about doctor recommendation in a mobile Medical Social Network (*m*-MSN) have not been explored yet.

5   Conclusion and Future Work

In this paper, we try to systematically investigate the problem of doctor recommendation, and made the experiments on the real-world dataset of a *m*-MSN. Experimental results show that we obtain the good accuracies of mining doctor-patient relationships from the network, the performance of doctor recommendation is also better than the baseline algorithms. As the future work, it would be interesting to combine individual information with random walk model. It is also interesting, from node mobility, to design location-based recommendation model.

Acknowledgements. This work was supported by the Natural Science Foun-dation of China (NSFC) under Grant No. 61272466, the National Natural Science Foundation of China (NSFC) under Grant No. 61303130 and the National Natural Science Foundation of China (NSFC) under Grant No. 61303233.

References

1. Wang, C., Han, J.W., Jia, Y.T., Tang, J., Zhang, D., Yu, Y.T., Guo, J.Y.: Mining advisor-advisee relationships from research publication networks. In: Proc. of SIGKDD, pp. 203–212 (2010)
2. Gong, J.B., Lu, S.L., Wang, R., Cui, L.: PDhms: pulse diagnosis via wearable healthcare sensor network. In: Proc. of ICC, pp. 1–5 (2011)
3. Gong, J., Sun, S.: Individual doctor recommendation model on medical social network. In: Tang, J., King, I., Chen, L., Wang, J. (eds.) ADMA 2011, Part II. LNCS (LNAI), vol. 7121, pp. 69–81. Springer, Heidelberg (2011)
4. Macdonald, C., Ounis, I.: Voting for candidates: adapting data fusion techniques for an Expert Search Task. In: Proc. of CIKM, pp. 387–396 (2006)
5. Tang, J., Sun, J.M., Wang, C., Yang, Z.: Social influence analysis in large-scale networks. In: Proc. of SIGKDD, pp. 807–816 (2009)
6. Yang, Z., Tang, J., Wang, B., Guo, J.Y., Li, J.Z., Chen, S.C.: Expert2Bole: from expert finding to Bole search (demo paper). In: Proc. of SIGKDD, pp. 1–4 (2009)
7. Yang, X.W., Steck, H., Liu, Y.: Circle-based recommendation in online social networks. In: Proc. of SIGKDD, pp. 1267–1275 (2012)
8. Cheung, K.W., Tian, L.F.: Learning user similarity and rating style for collaborative recommendation. Information Retrieval 7(3-4), 395–410 (2004)

9. Shenye, L., Jin, R.M.: Learning personal + social latent factor model for social recommendation. In: Proc. of SIGKDD, pp. 1303–1311 (2012)
10. Salakhutdinov, R., Mnih, A., Hinton, G.: Restricted Boltzmann machines for collaborative filtering. In: Proc. of ICML, pp. 791–798 (2007)
11. Tang, J., Wu, S., Sun, J.M., Su, H.: Cross-domain collaboration recommendation. In: Proc. of KDD, pp. 1285–1293 (2012)
12. Küçüktunç, O., Saule, E., Kaya, K., Çatalyürek, Ü.V.: Diversifying citation recommendations. Information Retrieval, 1–9 (2012)
13. Tang, J., Jin, R.M., Zhang, J.: A topic modeling approach and its integration into the random walk framework for academic search. In: Proc. of ICDM, pp. 1055–1060 (2008)
14. Feng, W., Wang, J.Y.: Incorporating heterogeneous information for personalized tag recommendation in social tagging systems. In: Proc. of SIGKDD, pp. 1276–1284 (2012)
15. Lee, Y.J., Huang, S.Y.: Reduced support vector machines: A statistical theory. IEEE Trans. on Neural Networks 18(1), 1–13 (2007)

From Post to Values: Mining Schwartz Values
of Individuals from Social Media

Mengshu Sun[1], Huaping Zhang[2], Yanping Zhao[3], and Jianyun Shang[1]

[1] School of Software Engineering, Beijing Institute of Technology, Beijing 100081, China
[2] School of Computer Science and Technology, Beijing Institute of Technology, Beijing, China
[3] School of Management and Economics, Beijing Institute of Technology, Beijing, China
kevinzhang@bit.edu.cn

Abstract. This paper aims to provide a novel method called Automatic Estimation of Schwartz Values (AESV) from social media, which automatically conducts text categorization based on Schwartz theory. AESV comprises three key components: training, feature extraction and values computation. Specifically, a training corpus is firstly built from the Web for each Schwartz value type and the feature vector is then extracted by using Chi statistics. Last but most important, as for individual values calculation, the personal posts are collected as input data which are converted to a word vector. The similarities between input vector and each value feature vector are used to calculate the individual value priorities. An experiment with 101 participants has been conducted, implying that AESV could obtain the competitive results, which are close to manually measurement by expert survey. In a further experiment, 92 users with different patterns on Sina weibo are tested, indicating that AESV algorithm is robust and could be widely applied in surveying the values for a huge amount of people, which is normally expensive and time-consuming in social science research. It is noted that our work is promising to automatically measure individual's values just using his/her posts on social media.

Keywords: Social Media, Schwartz values measurement, Text mining, AESV algorithms, Big data.

1   Introduction

Value is defined as "what is important to us in life", which reflects underlying social behaviors, personal behaviors and organizational behaviors [1]. There are various kinds of value models to represent "the values", among which Schwartz values model is proved to be universal among 82 countries [2]. Hundreds studies suggest that varied values priorities may result in different behaviors, altitudes or needs even in same context. According to these studies, a value analysis model can help to find out users' potential needs and predict their altitudes. Therefore it serves for business or political goals such as targeting marketing and public opinion monitoring in the domain of e-commerce or social media network. For example, in e-commerce, the type of products such as a car to be offered to a user may be different depending on his values with respect to hedonism or security.

H.-Y. Huang et al. (Eds.): SMP 2014, CCIS 489, pp. 206–219, 2014.

A considerable amount of research works have been done on Schwartz values model since 1992, most of which examine how this value model is related to various attitudes, opinions, behaviors, personalities and background characteristics [3].

Value measurement is discussed in most values related works. Most measure instruments of values, including SVS (Schwartz Values Survey) [4], PVQ (Portrait Value Questionnaire) [2] and PCVS (Pairwise Comparison Value Survey) [5] are based on Multi-Dimension Scale (MDS), which is proved to be reliable and valid by numbers of existing studies. Although some progress has been made in this area, at least two major obstacles must be overcome. Firstly, it is expensive and time-consuming to conduct a man-made survey. Secondly, the answers are subjective and the result quality is not stable, since it is highly associated with the survey and sampling procedure. In addition, the interviewees normally tend to hide their real intentions in the real survey. Our work tries to conduct such social survey or measurement on given population from different perspectives. It takes their posts on certain social media into consideration, such as Sina Weibo or twitter. Instead of survey, all the work can be completed on the Web automatically, which is therefore efficient and effective, and can be operated on millions of people on social networks.

Furthermore, research work on personality analysis based on individual text can be traced to a long time ago. Correlation between people's words use and their personality has been widely proved. Nevertheless, the lack of dataset becomes the bottleneck to such research before social media is widely used. However, UGC (user generated content) provide an easy way to access massive individual posts on social media. In this situation, we focus on values among various personality types. By utilizing the linguistic association between words use and values, an Automatic Estimation of Schwartz Values (AESV) method is proposed which builds a bridge between UGC. Furthermore, reliability and validity assessment demonstrate that AESV can be more efficient and effective than the classical Schwartz value survey methodology.

This paper is organized as follows. Background and related work is shown in section 2. Section 3 presents an elaborate description of AESV model. Section 4 demonstrates the experimental results for verifying the proposed model. Section 5 provides our research conclusions.

2 Background and Related Work

2.1 Values and Schwartz Values

Values, as a specific part of personality, are the socially desirable concepts used to represent goals mentally and the vocabulary used to express them in social interaction [1]. In recent studies and applications, values are widely applied in various disciplines, such as psychology, marketing and organizational behavior theory. Since values became a prominent topic according to Rokeach in 1968 [8], a large number of related studies have been conducted and several theories have been established over decades. There are two important viewpoints of values. First, values are crucial for

Table 1. Schwartz values

	Key phrase
Hedonism	pleasure, enjoying life
Achievement	successful, capable, ambitious, influential, intelligent, self-respect,
Power	social power, authority, wealth, preserving my public image, social recognition
Security	clean, national security, social order, family security, reciprocation of favors, healthy, sense of belonging
Tradition	devout, accepting portion in life, humble, moderate, respect for tradition, detachment
Conformity	politeness, honoring parents and elders, obedient, self-discipline
Stimulation	daring, a varied life, an exciting life
Benevolence	helpful, honest, forgiving, loyal, responsible, true friendship, a spiritual life, mature love, meaning in life
Universalism	protecting the environment, a world of beauty, unity with nature, broad-minded, social justice, wisdom, equality, a world at peace, inner harmony
Self-direction	creativity, curious, freedom, choosing own goals, independent

explaining social and personal organizational behaviors [2][6][7]. Secondly, values linked inextricably to affects [2], implying that enhancement of values arouses positive emotions and reduction of values brings about negative emotions. In recent studies, values conflict well explained the mechanism of mixed emotions which usually generated by controversial issues [10]. According to the two viewpoints, values are used to model and characterize cultural groups, societies, and individuals, to trace their behavior changes over a period of time, and to predict the future directions.

As a specific and typical value model, Schwartz values model comprises ten basic motivational values and 56 detailed values, as shown in Table 1 [3].

2.2   Values Analysis in Text

In decades, both the linguistic psychology and computational stylistics share a common theory that there are latent associations between personality and words use. According to linguistic psychology, since individuals vary in the words they use, it is reasonable to expect that this variation would reflect certain psychological differences [16]. A large number of correlations between words use and personality have been discovered by psychological studies since 1950s [11-17]. Each study concentrates on several specific parts of personality, including neuroticism, extraversion, self-esteem and so on.

The word counting method is still the favorite one in most studies that investigate the relationships between personality and words use. It involves predefining word categories of interest (e.g., emotion words, certainty words, and achievement words) and counting the proportion of words in a text sample that falls into those categories. There are three major interpretations for the situation [16]: (a) Word use may offer psychological information that cannot easily noted by the naked eye. (b) Usually, the psychological interpretation of a word count score is congruent with the definition of personality. (c) Growing evidences indicate that individual differences in written and spoken words use are stable across time and contexts.

Nevertheless, the lack of dataset is always a big problem in this kind of research until the period that big database came along with social network. Several studies have found the systematically personality and individual differences in large-scale corpus of posts. A research of personality in text is conducted on Google blog [18], which analyzes the correlation between Big Five personality scores that were evaluated by standard questionnaire and the count of words involved in LIWC2001 (Linguistic Inquiry and Word Count). Personality shows strong correlation both with aggregate words categories and single words. Since evidences between word use and personality have shown in a number of previous studies [15-17], researchers attempt to use words as a predictor of personality. Quercia et al. analyzes the relationship between Big Five personality and different types of Twitter users and propose a method to predict user's personality simply based on three counts which are following, follower and listed counts [19]. Gosling et al. reveal the relationship between Big Five and self-reported Facebook-related behaviors and observable profile information [20]. Moreover, machine learning is also utilized to predict personalities. Golbeck et al. trained two machine learning algorithms (M5'Rules and Gaussian Processes) to predict Big Five traits based on certain kinds of information from user's profiles [21]. In Minamikawa A et al., the estimation of Egogram, a specific part of personality, is performed using Multinomial Naïve Bayes classifier with features selected by information gain [22]. It is proved that feature words are strongly correlated with the characteristics of Egogram and moderate high accuracy can be reached. However, it's noteworthy that this method includes a training corpus of users who have been classified by answering a questionnaire, implying that if anybody wants to use this method in different social backgrounds especially in different languages, a complex and high-quality questionnaire is needed to prepared, which is time-consuming and expensive. This makes this method can hardly be applied to related research or practical intelligent services efficiently.

According to latest social network analysis report of IBM [23], though values along with BIG5 personality and needs can be estimated with 200 tweets, neither detailed methods nor result assessments have been provided. Research [24] attempts to detect the relationship between values and communication frequency on Enron Email. The word count method with a predefining word bag is used to estimate Schwartz values based on email text. However, this method is ineffective on public social network since most concept words in the word bag, such as universalism and hedonism, are rarely appeared. In addition, we cannot find any evidence, which can prove that the method used in that study can actually reflect Schwartz values. To the best of our knowledge, little work focuses on short, irregular and informal posts. Therefore, in this study, an automatic values estimation instrument is proposed by using text mining technique in big data environment. It is proved to be an alternative measure to multi-dimension scale used in classical study, according to the results of verification experiments.

3   AESV Model

The AESV (Automatic Estimation of Schwartz Values) model proposed in this paper is mainly based on the classification method with machine learning. It comprises three main parts as follows: **Generating feature index for value space**; **Calculating dynamic value weights vector; Estimating individual value priorities**.

3.1   Generate Features Index for Value Space

Generation of features index for value space is illustrated in this part. Top N features (the representative words) are selected from each value type. All features from 10 value types can be incorporated to a feature index with M dimensions, i.e., $F = [t_1, t_2, t_3, .., t_M]$.

The words bag of values used in prior study is usually composed of all the concept words, such as power, benevolence and hedonism. However, these concept words are less used in social media where people would like to talk about famous people or hot events related to values. For instance, people who talk about 'PM2.5' are likely to express their concerns about 'environment'. Therefore, an automatic extraction algorithm of key phrases is used in this paper. Currently, most existing extraction algorithms of key phrases are based on classified documents, which show several drawbacks when applied to social media. First, no classified documents can be used. Documents on Internet are always classified into practical categories such as sports, economic and politics, instead of concepts like universalism and tradition. Second, since the contents of social media normally depend on social background seriously, e.g., time and country, it is difficult to keep social media contents and key phrases extracted from static documents set synchronous. By considering this, news search engine is used to be a training data source and only titles and abstracts of news would be added into the corpus. News abstracts shares a lot of features with social media, including brief content, closely related to social background and updated quickly. Furthermore, since people talk about and repost news on social media, a lot of words circulating among social media come from news. These similarities combining a reasonable search strategy make Baidu news search result a dynamic corpus for values related words extraction.

In this stage, the first step is to build the corpus about these 10 value types. For the sake of the corpus quality, it's desirable to keep the query phrases comprehensive and less ambiguous. Therefore, all the 56 detailed values defined in Schwartz Values and their synonyms belonging to each value type are firstly used to build a 3-layer tree structure of query phrases, which is shown in Figure 1.

As a result, for each value type, an unique bag of key phrases can be built for news search on Baidu news engine. The query phrases comprise three parts: value type, detailed value and one of the synonyms. Since Baidu news engine ranks all news based on an aggregative indicator, which is compatible to people's attention, it's reasonable to simply crawl the abstracts of top 160 documents provided by Baidu news engine. Considering the number of synonyms word of detailed value v_i, the total search times is $\sum_1^{56} sc(v_i)$. and the number of news in each value types are $NN(vt_j) = \sum_{v_i \in vt_j} sc(v_i) \times 160$. Features extraction in this paper is based on the Pearson's chi-squared statistics ($\chi 2$).

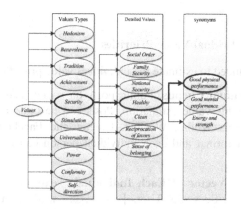

Fig. 1. The 3-layer tree structure of Search phrases

NOTE: The first column includes ten value types. The second column includes all 56 detail values defined by Schwartz [2], and the third column includes 198 synonyms derived from *Hownet* [25].

Three steps are performed to generate all feature words, which can be shown as follows.

Step 1: Word segmentation of all texts is used to discover new phrases, new symbols and meaningful strings with annotation by applying the Lexical analysis system ICTCLAS [26].

Step2: For each value category, counting the total occurrence frequency of each word and calculating the chi-square statistics to evaluate its significant relation to each category.

Step 3: Select the top 200 words in each value category, incorporate them to a whole feature index and delete any overlapping words that selected from different value categories.

3.2 Calculate Dynamic Value Weights Vector

Based on the M dimension feature index, each value type can be transformed to a specific vector, $Value_j=[w_{j1}, w_{j2}, .., w_{jM}]$, by calculating the weight of each feature corresponding to its category corpus. The calculation of the weight of each feature is conducted as a supervised learning task on the classified news based on Eq. (1).

$$Weight(w_j, C_i) = TF(w_j, C_i) \times \log\big(IDF(w_j)\big) \times IG(w_j) \tag{1}$$

TF(w_j, C_i) represents the feature frequency of word w_j in value type C_i, IDF(w_j) represents the inverse document frequency of a word w_j in whole corpus and IG(w_j) is the information gain of w_j. As a result, 10 vectors can be generated based on the feature index, which adapt to a certain period and the corresponding social background.

3.3   Estimate Individual Value Priorities

Three steps are included to estimate the individual value priorities.

(1) Building Individual Corpus. Public posts on social media are used to build a person's corpus for his value estimation. Posts can be collected via multiple social media, including Sina Weibo, WeChat, tweeter and Facebook. All posts can be transformed to text format and each individual person can be presented by his/her corpus.

(2) Calculating the Vector of Each Individual. Texts of each individual will be transformed to a specific vector User$_i$=[w_{i1}, w_{i2}, .., w_{iM}]. User value vector can be generated as the same way as the value type vector. Considering the value changing over time, only the latest 200 messages are used to compose the sample for each participant. All the 200 messages are gathered and treated as a whole document. Then, each document is transferred to a vector based on the feature index generated in feature selection step. Eq. (2), similar to Eq. (1), is used to calculate the weight on each feature of a user. In this way, each participant can be represented by a vector that concentrates on values along with the 10 value types. TF$\left(w_j, u_i\right)$ reprsents the feature frequency of a word w_j in user corpus u_i.

$$\text{Weight}\left(w_j, u_i\right) = \text{TF}\left(w_j, u_i\right) \times \log\left(\text{IDF}\left(w_j\right)\right) \times \text{IG}\left(w_j\right) \tag{2}$$

(3) Calculating the Similarity of User Vector and Each Value Vector. Value evaluation is based on a classical indicator, i.e., similarity, which describes the priority score of user value by two vectors' inner product, and can be calculated by following equation.

$$\text{Similarity}\left(u_i, C_j\right) = \cos\left(u_i, C_j\right) = \frac{u_i C_j}{|u_i||C_j|} \tag{3}$$

Ten similarity values can be obtained for each participant by this way, which represent their concern degrees on ten value types respectively. High similarity value implies high concern about this value type, and vice versa. For example, high similarity between a person vector and the universalism vector means that this person follows a lot in social media about environment protection, charity or nature beauty. It is assumed that people talk about things they care about; hence each similarity reflects how much he or she cares about each value type, which represents exactly his or her priority of values. In addition, the orders of them are considered to be the priorities of values.

4   Experiment

Two experiments have been performed to validate the advantage of our proposed method. Specifically, the first experiment aims to assess the reliability and validity of AESV, which are both used to investigate whether AESV can be an alternative instrument of three previous evaluation methods. In the second experiment, AESV model is employed to reveal the differences between the certificated users and the ordinary ones.

4.1   Data Collection

Datasets for the two experiments are described in **Dataset1** and **Dataset 2** as follows. All posts including posting and forwarding, are crawled from personal pages.

Dataset 1: This dataset is prepared for the experiment of validity and reliability assessment. Participants include 101 undergraduate and graduate students of Beijing Institute of Technology. By tracing these 101 user IDs on social networks, the latest 400 messages on social media (i.e., Sina Weibo, RenRen.com and QQ space) of each participant are collected to build his or her individual corpus. To perform the reliability assessment, each person's corpus would be divided into two parts with 200 messages in each by chronological order. Then AESV method is applied in these two parts of individual corpus separately. To conduct the validity assessment, four groups of value scores are required. Each participant is invited to answer 3 well-recognized scale questionnaires (i.e., PVQ, PCVS and Ranking) to get 3 groups of scores. The forth group is the first group of AESV scores.

Dataset 2: This dataset is prepared for the comparison analysis. In order to obtain samples as representative as possible, 92 individual participants who post over 200 messages on Sina Weibo are randomly selected with exclusion of the commercial ones, which includes 62 ordinary users and 30 certificated users. The latest 200 posts of each individual are scrapped from their personal Weibo pages. Thus, each user could obtain a group of scores of ten values estimated by AESV.

4.2   Experiments Results Analysis

As we described above, firstly, core features of Schwartz theory are examined in Experiment 1 (i.e., **Reliability and Validity Assessment Experiment**) on Dataset 1. Secondly, value pattern differences between disparate user groups on Sina Weibo are investigated by using AESV in Experiment 2 (i.e., **Comparison Analysis Experiment**).

Reliability and Validity Assessment
The proposed AESV model is expected to be an alternative measure instrument of Schwartz Values, which therefore should be acceptable to psychometric criterions compared with previous well-recognized measure instruments. Reliability and validity are two most essential criterions that can show the quality of a measure by assessing

the extent of random errors in survey response [27] and the score accuracy to reflect the theory concept.

(1) Reliability Assessment. Test-retest reliability avoids the problem of trying to produce parallel forms of a measure on two occasions and guarantees the stability of the measure. However, the intervals in test-retest reliability mostly keep in 10 to 30 days, which are inappropriate in AESV. If we derive 200 Weibo message from a user's page each time with an interval of 30 days, most messages would be repetitive. Therefore, we crawled the latest 400 unduplicated Sina Weibo messages of a user and divide them into two parts by chronological order. Two groups of value estimation are conducted separately based on these two parts, which are considered to be the test and the retest. Correlation coefficients are calculated between the first and second group estimations of AESV. The results of reliability (i.e., Spearman correlation coefficients) are shown in Table 2.

Table 2. Correlations between Schwartz 10 Value Types by AESV in Group 1 and Group 2

variable	Hedonism	Benevolece	Tradition	Stimulation	Security	Achievement	Universalism	Self-direction	Conformity	Power
Hedonism	.280**	-.077	.062	.187	-.009	.156	-.249*	-.201*	.100	-.134
Benevolece	.008	.773**	-.002	-.258**	-.165	-.166	.053	-.173	.137	-.385**
Tradition	-.102	-.128	.462**	.059	-.133	-.159	-.038	.105	.295**	-.234*
Stimulation	.213*	-.232*	.110	.545**	-.133	.048	-.162	-.008	-.136	-.122
Security	-.192	-.229*	-.039	-.078	.754**	-.213*	-.061	-.068	-.194	.052
Achievement	.030	-.114	-.446**	-.094	-.185	.537**	.049	-.012	-.120	.490**
Universalism	-.107	.291**	-.141	-.238*	-.024	-.176	.470**	-.199*	-.037	.009
Self-direction	-.100	-.188	-.011	-.077	.006	.057	-.077	.631**	-.265**	.037
Conformity	-.158	-.036	.214*	.096	-.098	-.132	-.018	-.049	.530**	-.218*
Power	.089	-.303**	-.170	-.038	-.134	.168	.032	.035	-.171	.603**

The within-value correlations (i.e., values on the diagonal) range from 0.280 to 0.773 with the characteristic that all these ten value types in two groups of estimations are strong correlated (p<0.01); While the cross-value correlations, ranging from 0.002 to 0.490, is relative low, implying that this measure is comparatively stable when randomly selecting 200 Weibo messages from the whole Weibo corpus of a person.

(2) Validity Assessment. Validity test is based on the Multitrait-Multimethod (MTMM) matrix [28]. The Pearson correlation coefficients are calculated between pair-wise comparisons of the four different methods on 10 value types (traits). Then the validity of convergence and discriminant can be evaluated by comparing the correlations among methods. There are two essential criteria [29] for convergence and discriminant validity as follows:

(a) Convergence: The monotrait-heteromethod coefficients (MHC) are significantly different from zero and sufficiently close to 1.

(b) Discriminant: MHC should be higher than the heterotrait-heteromethod (HHC).

Appendix presents the MTMM matrix. It obtained 600 Pearson correlation coefficients totally including 60 MHCs and 540 HHCs. The significance for the correlation coefficient test is marked with '*'and "**" as the strong (p-value<0.05) and very strong respectively (p-value<0.01).

To examine the two validity criteria above clearly, we summarize all important coefficients in MTMM into Table 3.

Table 3. Statistical results of MTMM

	PVQ/PCVS	PVQ/RANK	PVQ/AES\	PCVS/RANK	PCVS/AES\	RANK/AES\	Total
Count(MHC, p<0.05)	10	4	8	7	7	8	44
Count(MHC,p<0.01)	9	4	7	4	4	7	35
Min(MHC)	.241*	-0.1310	0.1541	0.0962	0.0321	0.0893	
Max(MHC)	.522**	.425**	.592**	.320**	.352**	.562**	
Min(HHC)	0.0014	0.0012	0.0000	0.0001	0.0000	-0.0003	
Max(HHC)	-.320**	-.363**	.475**	-.292**	-.340**	-.340**	
Count(HHC<Min(MHC))	84	72	64	55	16	47	338

Note: Table 3 is a summary of MTMM, e.g., 10 in R1C1 represents the number of significant MHCs between PVQ with PCVS. 9 in R2C1 represent the highly significant MHCs in the same pair. Min(MHC) and Max(MHC) (refer to MTMM for 0.241* (p<0.05) is the minimum of the coefficients between PVQ with PCVS) and maximum MHC between the two measure methods (0.286** (p<0.01)). Count(HHC<Min(MHC)) means the number of HHCs that is less than Min(MHC) between two measure methods.

Since it has been proven that the measurement of PVQ, PCVS and Rank can be substitution for each other, the count of MHCs and HHCs between them can be the baseline to evaluate the accuracy of AESV. In Table 3, 76.7% of MHCs meet the convergence validity criterion when p<0.05 (i.e., 44(R1C7) out of 60). Among the six groups of comparisons (combinations of each 2 of 4 measures), PVQ and PCVS show the strongest correlation. All (10(R1C1)) the correlation coefficients are significant (over 0.241*(R3C1)) and 90% of MHCs (9 (R2C1) out of all 10) are strong. From the results in Table3, AESV is pretty good to substitute for PVQ and RANK, comparing with PVQ/RANK, PCVS/RANK and PCVS/AESV. Moreover, 388 (R7C7) out of 540 comparisons meet the discriminant criterion, indicating clear discriminant validity for 10 value types. Therefore, both MHC and HHC show the relative high validity of Schwartz values measured by AESV.

In summary, both reliability and validity assessment present strong evidence that AESV can be an effective alternative instrument to estimate Schwartz values.

Comparison of Two Groups by AESV

One of the most important purposes of proposing AESV is to provide an easy way to conduct large-scale value-based analysis automatically on social networks. In this experiment, we simply use AESV to compare the value differences between certificated users and ordinary users. As shown in Table 4, ten pair-wise t tests are conducted to examine the value differences between certificated users (V-users) who are famous or have high social status and ordinary users (N-users). Significant differences are found in values of 'hedonism', 'benevolence', 'universalism' and 'power', which accords with the corresponding phenomenon on Sina Weibo. Ordinary users show more concerns about enjoying life, family and friends; therefore they obtain relative high scores on 'benevolence' and 'hedonism'. On the contrary, certificated users express more concerns on social events, such as ethic issues, policy, environment or charity, which results in high scores on universalism and power.

Table 4. Values differences between two user groups

values	AVE(N-users)	AVE(V-users)	P-value
Hedonism	0.6779	-0.3142	0.00002
Benevolence	1.6939	1.1008	0.00182
Tradition	-0.1659	-0.41559	0.08917
Stimulation	-0.2315	-0.5968	0.11363
Security	-0.4744	-0.2835	0.19644
Achievement	0.1363	0.2127	0.63311
Universalism	-0.2108	0.4339	0.00012
Self-directic	-1.0136	-0.6988	0.07568
Conformity	-0.237	-0.0329	0.10408
Power	-0.1748	0.5944	0.00007

5   Discussions and Conclusions

In this paper, our contributions lie in:

(a) Providing a high-quality and low-cost evaluation instrument of Schwartz values based on social media;

(b) Verifying this method by using reliability and validity criteria;

(c) Investigating value differences between certificated and ordinary groups in a real case by taking advantage of the benefits of the measure instrument proposed in this study.

Firstly, we have described the process of conducting AESV instrument to estimate Schwartz Values of individuals, including corpus collection, feature selection, classified training and values estimation. Secondly, we have investigated the validity and reliability of AESV. Both two criteria indicate that individual values can be deduced based on his or her posts on social networks. Thirdly, by using AESV, different values patterns have discovered among people with various social roles. This experiment exactly shows the most meaningful advantage of this AESV, that is we can estimate personal values automatically without the participation of each person, which means that large-scale values analysis of persons or groups can be proceed based on this AESV easily by any researchers or service/commodity providers.

Instead of using a sampling and questionnaire method, AESV can estimate values pattern of a group or individual effectively. Therefore it can provide personalized information for service/commodity provider and especially contribute to target marketing, user interface design and recommendation systems. Furthermore, since strong correlations between values and behaviors or attitudes have been proved in previous studies, AESV can provide meaningful references to predict sentiment trends and assist public opinion analysis for policy makers or researchers.

Acknowledgments. This work is partly supported by the National Science Foundation of China (No. 61272362) and the Fundamental Research Funds for the Central Universities (No. 2014RC0601).

References

1. Mayton, D.M., Furnham, A.: Value Underpinnings of Antinuclear Political Activism: A Cross-National Study. Journal of Social Issues 50(4), 117–128 (1994)
2. Schwartz, S.H.: An Overview of the Schwartz Theory of Basic Values. Online Readings in Psychology and Culture 2(1), Article 11 (2012)
3. Schwartz, S.H., Cieciuch, J., Vecchione, M., Davidov, E., Fischer, R., Beierlein, C., Konty, M.: Refining the theory of basic individual values. Journal of Personality and Social Psychology 103(4), 663 (2012)
4. Schwartz, S.H.: Universals in the content and structure of values: Theory and empirical tests in 20 countries. In: Zanna, M. (ed.) Advances in Experimental Social Psychology, vol. 25, pp. 1–65 (1992)
5. Schwartz, S.H.: Are there universal aspects in the structure and contents of human values? Journal of Social Issues 50(4), 19–45 (1994)
6. Durkheim, E.: Suicide. Glencoe. Understanding suicidal behavior: The suicidal process approach to research and treatment (1951)
7. Weber, M., Parsons, T.: The Protestant ethic and the spirit of capitalism. Roxbury Pub. (1998)
8. Rokeach, M.: Beliefs, attitudes and values: A theory of organization and change (1968)
9. Sagiv, L., Schwartz, S.H.: Value priorities and subjective well-being: Direct relations and congruity effects. European Journal of Social Psychology 30(2), 177–198 (2000)
10. Hanselmann, M., Tanner, C.: Taboos and conflicts in decision making: Sacred values, decision difficulty, and emotions. Judgment and Decision Making 3(1), 51–63 (2008)
11. Allport, G.: Pattern and growth in personality. Holt, Rinehart, & Winston, New York (1961)
12. Allport, F., Walker, L., Lathers, E.: Written composition and characteristics of personality. Archives of Psychology 173, 1–82 (1934)
13. Berry, D.S., Pennebaker, J.W., Mueller, J.S., Hiller, W.: Linguistic bases of social perception. Personality and Social Psychology Bulletin 23, 526–538 (1997)
14. Mehl, M.R., Pennebaker, J.W.: The social dynamics of a cultural upheaval: Social interactions surrounding September 11. Psychological Science 14(6), 579–585 (2003)
15. Oberlander, J., Gill, A.J.: Language with character: A corpus-based study of individual differences in e-mail communication. Discourse Processes 42(3), 239–270 (2006)
16. Fast, L.A., Funder, D.C.: Personality as manifest in word use: Correlations with self-report, acquaintance report, and behavior. Journal of Personality and Social Psychology 94(2), 334 (2008)
17. Holtgraves, T.: Text messaging, personality, and the social context. Journal of Research in Personality 45(1), 92–99 (2011)
18. Yarkoni, T.: Personality in 100,000 words: A large-scale analysis of personality and word use among bloggers. Journal of Research in Personality 44(3), 363–373 (2010)
19. Quercia, D., Kosinski, M., Stillwell, D., Crowcroft, J.: Our Twitter Profiles, Our Selves: Predicting Personality with Twitter. In: 2011 IEEE Third International Conference on Privacy, Security, Risk and Trust (PASSAT) and 2011 IEEE Third International Conference on Social Computing (SocialCom), pp. 180–185, 9–11 (2011)
20. Gosling, S.D., Augustine, A.A., Vazire, S., Holtzman, N., Gaddis, S.: Manifestations of personality in online social networks: Self-reported Facebook-related behaviors and observable profile information. Cyberpsychology, Behavior, and Social Networking 14(9), 483–488 (2011)

21. Golbeck, J., Robles, C., Turner, K.: Predicting personality with social media. In: CHI 2011 Extended Abstracts on Human Factors in Computing Systems, pp. 253–262. ACM (2011)
22. Minamikawa, A., Yokoyama, H.: Blog tells what kind of personality you have: egogram estimation from Japanese weblog. In: Proceedings of the ACM 2011 Conference on Computer Supported Cooperative Work, pp. 217–220. ACM (2011)
23. Kelley, M.: IBM Researcher can Build a Detailed Personality Profile of you Based on 200 Tweets. Business Insider (2013), http://www.businessinsider.com/
24. Zhou, Y., Fleischmann, K.R., Wallace, W.A.: Automatic text analysis of values in the enron email dataset: Clustering a social network using the value patterns of actors. In: 2010 43rd Hawaii International Conference on System Sciences (HICSS), pp. 1–10. IEEE (2010)
25. Zhendong, D., Qiang, D.: HowNet (1999)
26. Zhang, H.-P., et al.: An Extraction method and system of terminology: China. 200710121839.0 (2007)
27. Lord, F.M., Novick, M.R.: Statistical Theories of Mental Test Scores. Addison-Wesley Publishing Company, Inc. (1968)
28. Campbell, D.T., Fiske, D.W.: Convergent and discriminant validation by the Multitrait-Multimethod matrix. Psychological Bulletin 56(2), 81 (1959)
29. Oishi, S., Schimmack, U., Diener, E., et al.: The measurement of values and individualism-collectivism. Personality and Social Psychology Bulletin 24(11), 1177–1189 (1998)

6   Appendix

Coefficients for Pair-wise Test on Convergent and Discriminant Validity for 4 Evaluation Methods

variable	methods	Hed	Ben	Tra	Sti	Sec	Ach	Uni	Sel	Con	Pow
Hedonism	PVQ/PCVS	.458**	-0.1379	0.0365	0.1762	-0.0905	-0.0946	-0.0287	-0.1162	0.0082	-0.0709
	PVQ/RANK	-0.1310	0.0191	0.0052	-0.0105	-0.0188	-0.1010	0.0162	.239*	0.0520	-0.0737
	PVQ/AESV	0.1592	0.0423	-0.1364	-0.0202	-0.0745	-0.0265	0.0419	0.0017	-0.1533	0.0881
	PCVS/RANK	.320**	0.1714	-0.1621	-0.0185	-0.0480	-0.1915	0.0673	.219*	-0.0678	-.292**
	PCVS/AESV	0.1649	-0.1814	0.0527	0.1900	-0.0974	-0.0694	-0.0465	-0.0314	-0.0437	0.1032
	RANK/AESV	0.1659	0.0625	-0.0397	-0.0883	-0.0095	-0.0049	-0.0950	-0.0361	0.0324	-0.0003
Benevolece	PVQ/PCVS	0.1021	.415**	-0.0548	-0.0944	-0.1154	-.233*	0.1359	0.0865	-.299**	-0.1472
	PVQ/RANK	0.0124	.296**	-0.0328	0.0650	-0.0382	-0.0786	-0.1773	-0.1134	-0.0707	0.0706
	PVQ/AESV	0.0425	.305**	-0.1323	-0.0322	-0.0451	0.0029	.199*	-.272**	-0.1022	-0.0925
	PCVS/RANK	-0.1869	.271**	-0.0088	-0.1325	-0.0003	-0.1551	-0.0062	-0.0276	0.0723	0.0660
	PCVS/AESV	0.0641	.345**	-0.1351	-0.0182	-0.1336	-0.0752	.228*	-0.1676	-0.0730	-0.1698
	RANK/AESV	0.0118	.430**	0.1607	-0.1939	-0.1858	-0.1204	0.1068	-.243*	0.1386	-0.1751
Tradition	PVQ/PCVS	-0.1090	0.0574	.241*	0.0134	0.0875	-0.0606	-0.1511	-0.1490	0.1571	-0.0178
	PVQ/RANK	0.0316	-0.0892	.425**	.283**	-0.0521	-.203*	-.209*	-0.0928	0.1274	0.0054
	PVQ/AESV	-0.0732	0.1309	.592**	-0.0667	-0.1672	-.272**	-0.1212	0.0140	.341**	-.271**
	PCVS/RANK	-0.0499	-0.0593	0.1869	0.0627	0.0768	-0.0843	-0.0069	0.0700	0.0410	-0.1504
	PCVS/AESV	-0.1448	0.0008	0.1524	-0.1595	.237*	-0.0997	-0.0093	-0.1889	.204*	0.0323
	RANK/AESV	-.216*	-.277**	.562**	0.1141	0.0521	-.218*	-0.1059	0.0559	0.1007	0.0735
Stimulation	PVQ/PCVS	0.0014	0.0194	0.1038	.286**	-.320**	-0.0280	-0.1748	0.0564	0.1138	0.1430
	PVQ/RANK	0.1100	-0.0799	0.0215	0.1174	-0.1840	-0.0052	0.0481	0.0333	-0.1284	0.1141
	PVQ/AESV	0.1427	-.318**	.475**	.338**	-0.0573	-.254*	-0.1693	0.0902	0.1431	-.227*
	PCVS/RANK	0.0013	-0.0410	-0.1535	.268**	-.276**	0.0959	-0.0480	-0.0080	-0.0291	0.1388
	PCVS/AESV	0.1499	-0.1379	0.1436	.352**	-0.1553	-0.0145	-.216*	0.0343	0.0327	-0.0990
	RANK/AESV	.210*	-0.1531	-.214*	.355**	-0.1833	0.1097	-0.0885	0.1953	-0.0599	-0.1204
Security	PVQ/PCVS	-.288**	-0.1826	-0.0067	-0.1741	.490**	0.1079	0.0482	-0.0901	0.0573	-0.0701
	PVQ/RANK	-0.0070	0.0012	.296**	-0.0515	0.1642	-0.1892	0.1258	-0.0801	0.0586	-.283**
	PVQ/AESV	-.201*	-0.0042	-0.0504	-0.1141	.489**	-0.1719	-0.1312	0.0699	-0.0207	0.0298
	PCVS/RANK	-0.1411	-0.1331	.247*	-0.0934	.230*	-0.0681	-0.0463	-0.0323	0.1663	-0.0308
	PCVS/AESV	-.340**	-0.0392	0.1083	-0.1127	.285**	-0.1172	0.0124	0.0627	0.0126	0.1028
	RANK/AESV	-0.0613	-0.0920	-0.0186	-0.1791	.473**	-0.0860	-0.0088	-0.0253	0.0382	-0.0784
Achieveme nt	PVQ/PCVS	-0.0048	-.274**	-.252*	-0.0902	.202*	.317**	0.1060	-0.1033	-0.1204	0.1263
	PVQ/RANK	-0.0718	-0.0099	-.363**	-.216*	-0.0686	.375**	0.0855	-0.0031	-0.1952	.281**
	PVQ/AESV	0.0099	-0.1633	-.223*	-0.1417	-0.0936	.376**	-0.0322	0.0295	-0.0859	.390**
	PCVS/RANK	0.1140	-0.1739	-0.1243	0.0887	-0.0652	.314**	0.0062	-0.0643	-0.1936	0.1037
	PCVS/AESV	-0.1036	-.208*	-0.1762	0.0539	0.0442	.235*	-0.0832	.235*	-0.1860	.217*
	RANK/AESV	0.0615	-0.0784	-.299**	-0.0503	-0.0581	.497**	0.0578	-0.1572	-.212*	.263**
Universalism	PVQ/PCVS	0.1020	0.1141	-0.0861	-0.1135	0.0388	0.0572	.341**	-0.1687	-0.1810	-.206*
	PVQ/RANK	-0.0245	0.1149	-0.1406	-0.0871	0.0271	0.0856	0.0445	0.0508	-0.1083	-0.0791
	PVQ/AESV	-0.1419	0.1885	-.274**	-0.0967	-0.0457	-0.0254	.376**	0.0171	-0.1146	-0.0012
	PCVS/RANK	-0.0500	.207*	0.0232	-.252*	-0.0216	-0.1409	.212*	-.235*	0.0963	0.0706
	PCVS/AESV	-0.1170	0.0773	-0.1875	-0.0395	-0.0063	-0.0446	.199*	-0.0022	-0.1655	0.1814
	RANK/AESV	-0.0636	.274**	-0.0892	-.340**	-0.0771	-0.0712	.449**	-0.1302	-0.1146	0.0317
Self-direction	PVQ/PCVS	-0.1581	-0.1277	0.0420	0.0586	0.0665	-0.1201	-.274**	.522**	0.0316	-0.0582
	PVQ/RANK	0.0067	-.255*	-0.0504	0.0598	0.1095	0.0372	0.0771	0.0045	-0.0035	0.0516
	PVQ/AESV	0.0228	-0.1025	-0.0262	0.0000	0.0182	0.0205	-0.0475	.235*	-0.0424	-0.0579
	PCVS/RANK	0.0001	-.234*	-0.1805	0.0810	0.0776	0.1888	0.0527	0.1731	-0.1590	-0.0034
	PCVS/AESV	0.1281	-0.0261	0.1237	-0.0484	-0.0163	0.0041	-.230*	.207*	0.1085	-0.1928
	RANK/AESV	-0.0777	-0.1800	0.0365	.316**	-0.0581	-0.0573	-.270**	.502**	-0.0891	-0.0828
Conformity	PVQ/PCVS	0.0595	0.1616	0.1771	0.0353	-0.1506	-0.0876	-0.1928	-0.0908	.357**	-0.0858
	PVQ/RANK	0.0952	0.0620	-0.0579	0.0620	0.0358	-0.0978	-0.1810	-0.0606	.325**	-0.0638
	PVQ/AESV	-0.0964	-0.1287	.412**	0.0329	0.0935	-.266**	-0.1449	0.0127	.403**	-0.1746
	PCVS/RANK	-0.1767	-0.0755	0.0846	0.0742	-0.0193	0.0468	-0.1192	0.0630	.245*	-0.0728
	PCVS/AESV	0.0889	0.1003	0.0436	-.198*	0.0736	-0.0632	0.0675	-.201*	.286**	-0.1671
	RANK/AESV	0.1354	-0.0869	-0.0210	0.1337	0.0740	-0.1276	-0.1129	0.0189	0.0893	-0.0911
Power	PVQ/PCVS	-0.1369	-0.1278	-0.0908	-0.0235	-0.1437	0.1444	0.0718	-0.0453	0.0229	.362**
	PVQ/RANK	-0.0187	-0.0313	-0.1165	-.203*	-0.0143	0.1057	0.1468	0.0243	-0.0308	0.0568
	PVQ/AESV	0.1595	-0.1207	-.207*	.225*	-.226*	.411**	-0.0790	-0.1351	-0.0932	0.1541
	PCVS/RANK	0.1410	0.0182	0.0169	-0.0647	0.0732	0.0110	-0.0329	-0.1106	-0.1364	0.0962
	PCVS/AESV	0.0594	-0.0433	-0.0135	0.0047	-0.1196	0.1793	0.0000	0.0198	-0.0790	0.0321
	RANK/AESV	-0.1448	0.0077	-0.1038	-0.0467	-0.0023	0.0863	0.0668	-0.0914	0.0093	.217*

Personality Prediction Based on All Characters of User Social Media Information

Danlin Wan, Chuang Zhang, Ming Wu, and Zhixiang An

Beijing University of Posts and Telecommunications, Beijing 100876, China
{wandanlin2014,azx-c}@163.com,
{zhangchuang,wuming}@bupt.edu.cn,

Abstract. In recent years, the number of social networks users has shown explosive growth. In this context, social media provides researchers with plenty of information about user behavior and social behavior. We are beginning to understand user's behavior on social media is related to user's personality. Conventional personality assessment depends on self-report inventory, which costs a lot to collect information. This paper tries to predict user's Big-Five personality through their information on social networks. We conducted a Big-Five personality inventory test with 131 users of Chinese social network Sina Weibo, and crawled all of their Weibo texts and profile information. By studying the relevance between all types of user generated information and personality results of users, we extracted five most relative dimensionalities and used machine learning method to successfully predict the Big-Five personality of users.

Keywords: Social media, Personality prediction, Machine learning.

1 Introduction

Social media is the social interaction among people in which they create, share or exchange information and ideas in virtual communities and networks. Nowadays, people use more and more time in social media, every time there is a flood of information creation and dissemination. These information includes user's social behavior, user generated texts and language habit, to some extent reflects user's personality.

Personality uniquely characterizes an individual, and profoundly influences user's mental status and social behaviors [1]. Conventional personality assessments use self-reported inventory [2]. Self-reported method has a profound theoretical significance, but it costs long time and lots of manpower.

This paper tries to use user generated information on social network which is easy to get to predict user's personality.

First core research point in this paper is why user generated information on social network can be used to predict user's personality, in other words, the relationship between information on social network and user's personality. Previous work has shown that the information in users' Facebook profiles is reflective of their actual personalities, not an "idealized" version of themselves [3]. Gosling et al. [10]

H.-Y. Huang et al. (Eds.): SMP 2014, CCIS 489, pp. 220–230, 2014.

delivered a mapping between 11 features of users' online behaviors on Facebook and users' personality. They verified there is a correlation between them.

Second core research point in this paper is how to use user generated information to predict user's personality. Most recent research tends to use one aspect of information on social network to predict user's personality. Globeck tried to predict web users' personality traits through text features on Facebook and Twitter [5], [6]. Quercia proposed to predict web users' personality traits through three features (i.e., following, followers and listed counts) available on profiles of Twitter [7]. Shuotian Bai and Bibo Hao used multi-task regression method to predict user's personality with their behavioral characteristics on Sina Weibo [8].

In this paper, we are interested in Chinese social media. Sina Weibo is one of the most popular sites in China. Akin to a hybrid of Twitter and Facebook, it is in use by well over 30% of Internet users, with a market penetration similar to the United States' Twitter [9]. Sina Weibo has been the leading micro blogging service provider in China. By the time March 2014, Sina Weibo had 143.8 million monthly active users, 66.6 million active users. So it's chose as the major source of user generated information.

In order to achieve these two main research cores. We conducted a Big-Five personality inventory test with 131 users of Sina Weibo, through result analysis, we got their Big-Five personality dimensions scores. We take these scores as ground truth. Then we crawled all of their Weibo texts and profile information including text features, social behavior features and interaction features. Through a Pearson correlation analysis between all features and user's Big-Five personality dimensions scores, we verified there is a correlation between them. According to the correlation results, we extracted the first five relative features and used logistic regression and Naive Bayes algorithms to learn and successfully predict user's personality.

In psychology, the Big Five personality traits are five broad domains or dimensions of personality that are used to describe human personality. The theory based on the Big Five factors is called the Five Factor Model (FFM) [11]. It was one of the most well-researched and well-regarded measures of personality structure in recent years. Tupes and Christal came up with five domains of personality, Openness, Conscientiousness, extroversion, Agreeableness, and Neuroticism first through analyses of previous personality tests [12], [13]. Latter research has proved that different tests, languages, and methods of analysis do not alter the model's validity [14], [15], [16]. Such comprehensive research has led to many psychologists to accept the Big Five as the current definitive model of personality [17], [18]. Following is a summary of the factors of the Big Five and their constituent traits.

- Openness to experience: Openness reflects the degree of intellectual curiosity, creativity and a preference for novelty and variety a person has.
- Conscientiousness: A tendency to be organized and dependable, show self-discipline, act dutifully, aim for achievement, and prefer planned rather than spontaneous behavior.

- Extraversion: Energy, positive emotions, surgency, assertiveness, sociability and the tendency to seek stimulation in the company of others, and talkativeness.

- Agreeableness: A tendency to be compassionate and cooperative rather than suspicious and antagonistic towards others. It is also a measure of one's trusting and helpful nature, and whether a person is generally well tempered or not.

- Neuroticism: The tendency to experience unpleasant emotions easily, such as anger, anxiety, depression, and vulnerability. Neuroticism also refers to the degree of emotional stability and impulse control and is sometimes referred to by its low pole, "emotional stability".

2   Related Studies

In the area of predicting user's personality according to their online behavior of social network, most of foreign research is based on Facebook or Twitter.

Globeck shown that users' Big Five personality traits can be predicted from the public information they share on Facebook. They collected fifty users' text information on Facebook and their Big-Five personality results, then used two regression algorithms ZeroR and Gaussian Processes to predict scores on each of the five personality traits [5], [6]. They can predict scores on each of the five personality traits to within 11% - 18% of their actual value.

Shuotian Bai and Bibo Hao used multi-task regression method to predict user's personality with their behavioral characteristics on Sina Weibo [8]. They examined the personality inventory test of 444 users, and extracted users' profile information and interaction information. Then used Muti-task regression and incremental regression to predict the Big-Five personality.

All in all, the current researches on this area are more concentrated on only one type of information on social network, such as text features or profile features. So they can't reveal the relationship between all types of information and personality dimensions. In this paper, we choose most benefit features for each personality dimensions, and successfully predict user's personality.

3   Method

The overall frame work in our research is shown in Figure 1. It shows the main procedures to collect user's personality and Social network information and how we use it to predict user's personality.

3.1   Participants

The survey was conducted during March 2014 to May 2014. A total of 589 individuals participated in our test, and 131 participants were recruited. Because this research defines qualified participants as who has more than 200 Weibo statuses and whose answer time was longer than 100 seconds.

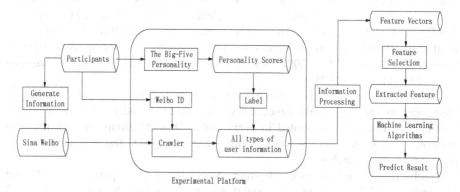

Fig. 1.Frame work of personality prediction

3.2   The Big-Five Personality Scores

Using standard Big-Five personality test scale, we conducted a personality test of Sina Weibo users. Standard Big-Five scale contains of 60 multiple-choice questions, each question on one dimension for evaluation. Options for each question are strongly agree, agree, not sure, do not agree, strongly disagree, to show measured by the bias in the measured dimension.

Personality values for each dimension range from -24 to 24. The larger positive value indicates more positive related, the larger negative value indicates more negative related, closer to 0 indicate that the characteristic dimensions is not obvious. Each dimension was divided to -24 to -8, -8 to 8, 8 to 24, these three levels representing a negative prominent, not obvious and positive prominent. In the following, we will use -1, 0, and 1 to represent these three levels.

3.3   Features Extracted from User Generated Information

In order to obtain user's generated information on Sina Weibo, we used crawler to get all of the 131 participants' profile information and Weibo text information.

All types of information can be categorized into 3 groups. The first group is user behavior which contains Weibo statues count, followers count, followings count, time since registration Sina Weibo. These information reflects user's basic use condition of Sina Weibo. The second group is interaction behavior which extracted from text information and contains expressions count, topics count and @ mentions. The third group is text features which reflect user's language habits on social network.

For the purpose of analyzing the content of Weibo content, we try to use the Linguistic Inquiry and Word Count (LIWC) dictionary [19]. LIWC produces statistics on 71 different features of text in five categories. These include Standard Counts (word count, words longer than six letters, number of prepositions, etc.), Psychological

Processes (emotional, cognitive, sensory, and social processes), Relativity (words about time, the past, the future), Personal Concerns (such as occupation, financial issues, health), and Other dimensions (counts of various types of punctuation, swear words). We can use this dictionary to catch characteristics of the text. So we choose LIWC2007 Simplified Chinese Dictionary. Firstly, using IKanalyzer tool (a Chinese word segmentation tool) to get word segmentation results. Then bring word segmentation results to match LIWC2007 Simplified Chinese Dictionary, and get all mapping count of each feature of LIWC.

The second group and third group are both extracted from text information, we need some processes to obtain standard text features. In second group, three features' average frequency was computed in every status. In third group, through dividing mapping count by user's total words count, we get 71 LIWC features' frequency in per Weibo status of each user.

Finally, through information collecting, processing and computing, we collected 131 participants' 5 dimensions personality scores, 4 features about user behavior, 3 features about interaction behavior and 71 text features.

3.4   Feature Selection

Considering the size of the data set and the numbers of extracted features, we can't use all these features to predict user's personality. So we need to select features. In this paper, we use Pearson Correlation Coefficient as our standard to select features based on dependency metrics theory. Besides, Pearson Correlation Coefficient can reveal the relationship between all types of information and user's personality.

Pearson correlation coefficient describes the degree of tightness between two fixed variables and defined as

$$p = \frac{\Sigma(X_i - \bar{X})(Y_i - \bar{Y})}{\sqrt{\Sigma(X_i - \bar{X})^2}\sqrt{\Sigma(Y_i - \bar{Y})^2}} \tag{1}$$

In the use of this formula, we replaced X_i with extracted features values and replaced Y_i with different personality dimensionalities scores. And $\bar{X}$ and $\bar{Y}$ representative the mean values of them.

The Pearson Correlation value ranges from -1 to +1. If it is positive, the two variables are positively correlated (the greater the value of one variable, the greater the value of another variable). If it is negative, the two variables are inversely related (the smaller the value of a variable, the greater the value of another variable).

We firstly analyzed the Pearson Correlation Coefficient between 3 groups' information and personality scores. The results are show in Table 1, Table 2 and Table 3. Because text information has 71 features, so we only choose highly relevant characteristics to show.

Table 1. Pearson Correlation values between personality scores and interaction behavior. 3 features of interaction behavior mean frequency in every status. Bold font indicates a relatively high degree of correlation.

	Expressions	Topic	@
Neur.	-0.012	0.122	0.057
Extr.	**0.172**	0.041	**0.148**
Open.	**-0.16**	-0.01	-0.084
Agree.	**0.161**	0.032	-0.045
Cons.	-0.041	-0.085	-0.008

Table 2. Pearson Correlation values between personality scores and user behavior. Statues means user's statues count, followers and followings mean user's followers or followings count. Time means time since registration Sina Weibo. Bold font indicates a relatively high degree of correlation.

	Statues	Followers	Followings	Time
Neur.	**-0.124**	-0.085	0.107	-0.042
Extr.	0.012	-0.011	0.02	**0.158**
Open.	-0.112	-0.091	-0.071	**-0.145**
Agree.	0.068	**0.15**	0.11	**0.173**
Cons.	0.066	0.02	**-0.127**	0.091

Table 3. Pearson Correlation values between personality scores and text features. Bold font indicates a relatively high degree of correlation.

	anger	sexual	certain	body	ProgM	home
Neur.	**0.23**	**0.235**	0.203	**-0.25**	**-0.204**	-0.201
	nonfl	MultiFun	time	swear	anger	sexual
Extr.	**0.178**	**0.152**	0.137	**-0.165**	**-0.144**	-0.138
	space	work	motion	see	humans	funct
Open.	0.191	0.165	0.132	**-0.348**	**-0.277**	**-0.24**
	feel	discrep	body	space	motion	relativ
Agree.	**0.242**	**0.231**	**0.174**	**-0.183**	-0.172	-0.17
	ProgM	feel	motion	funct	i	tentat
Cons.	0.155	0.132	0.111	**-0.222**	**-0.193**	**-0.192**

Through these three tables, we can see that different features are related to different personality dimensionalities. And not all features are related to personality. In interaction behavior group, expressions show relatively high correlation with 3 personality dimensionalities. In user behavior group, time shows relatively high correlation with 3 personality dimensionalities. And text group has most features that have relatively high correlation with personality dimensionalities. And most of the relatively high correlation values are range from 0.17 to 0.35. These values mean there are relations between user generated information on Social media and personality and the correlation intensity is weak correlation. The results correspond to the research on Twitter [5].

Many of the correlations make intuitive sense. For example, neuroticism is positively correlated with words about anger (e.g. "fury", "rage"), suggesting neurotic people tend to express more anger emotion on social media. At the same time, the words about anger are negatively correlated with extraversion, indicating extraverted people tend to express less about anger emotion. Extraverted people also tend to use more expressions and @ mentions in their Weibo statues.

In order to get better predict result, we selected five features for each personality dimensionality to predict user's personality. The criterion is choosing high Pearson Correlation values features, at the same time, every personality dimensionality should have positive related features and negative related features. The eventual result shows in Table 4. All three types of features are used, it means different features match different personality dimensionalities. So we can't use only one type features to predict all five personality dimensionalities.

Table 4. Top five high Pearson Correlation values between personality scores and all types' features of every personality dimensionalities. Bold font indicates features except text features.

	sexual	anger	certain	body	ProgM
Neur.	0.24	0.23	0.203	-0.25	-0.204
	swear	anger	nonfl	**Expressions**	**Time**
Extr.	-0.2	-0.144	0.178	0.172	0.158
	space	work	see	funct	humans
Open.	0.19	0.165	-0.346	-0.24	-0.277
	feel	discrep	space	**Time**	motion
Agree.	0.24	0.231	-0.183	0.173	-0.172
	ProgM	feel	funct	i	tentat
Cons.	0.16	0.132	-0.222	-0.193	-0.192

3.5   Personality Prediction

In order to predict the scores of five personality dimensionalities, we divided our dataset into training set and test set and performed machine leaning test on Weka [20], which is a software of machine learning and data mining. Using extracted five features for different personality dimensionalities to predict personality. The predicting outcomes for one user are like the Fig. 2. Each user has five scores for five personality dimensionalities.

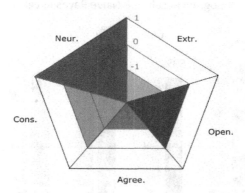

Fig. 2. Predicting outcomes for one user

We used two algorithms: Logistic Regression and Naïve Bayes to training data, each algorithm with a 10-fold cross-validation with 10 iterations. Then used them to compute personality results of test set, and used personality test result which is considered as ground truth in this paper to verify the predicted results. The precision and recall of each algorithm is shown in Fig. 2 and Fig. 3.

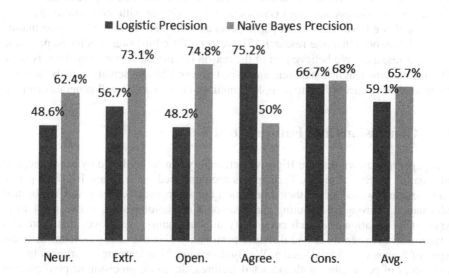

Fig. 3. Precision for each algorithm and personality dimensionalities

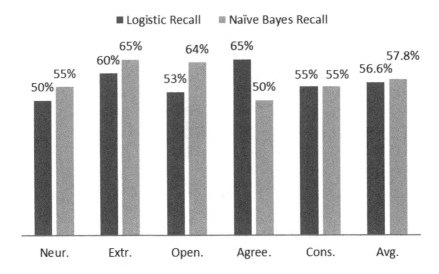

Fig. 4. Recall for each algorithm and personality dimensionalities

We can find that in general Naïve Bayes has better performance on precision, and two algorithms have similar performance on recall. Considering best result of two algorithms, we can find that neuroticism was the most difficult to predict and Openness and agreeableness are easy to compute. This is similar with the result in [5].

Although we have much more participants and consider more types of information on social networks than the research [5], we get a little bit lower prediction precision than their research. We believe part of the reason is Chinese semantic analysis is more difficult than the English semantic analysis, because Chinese semantic analysis needs to segment words according to complex language custom and it will generate error.

4   Conclusions and Future Work

This paper has shown that the Big-Five personality can be predicted by user generated information on Social media. Participants accomplished the standard Big-Five personality test and we collected their interaction behavior, user behavior and Weibo text information. Through computing the Pearson Correlation values between all three types of information and each personality dimensionality, we proves that there is a relationship between them and different personality dimensionality matches different types of features on Social media. We used two machine learning algorithms to predict scores of personality with extracted features, the mean precision of prediction of five personalities was 0.707.

In the future, we will continue recruit more participants and collect their information on Sina Weibo in order to get a larger dataset. And we will improve the algorithms to predict personality. Our main research direction is to reveal the relation between use's personality and their strong ties friends' personality on social network. And if there is a relation, we can use it to get a better performance to predict user's personality.

References

1. Goldberg, L.: The structure of phenotypic personality traits. American Psychologist 48(1), 26 (1993)
2. Back, M., Stopfer, J., Vazire, S., Gaddis, S., Schmukle, S., Egloff, B., Gosling, S.: Facebook Profiles Reflect Actual Personality, Not Self-Idealization. Psychological Science 21(3), 372 (2010)
3. Burger, J.: Personality. Thomson Wadsworth, Belmont (2008)
4. Li, L., Li, A., Hao, B., Guan, Z., Zhu, T.: Predicting Active Users' Personality Based on Micro-Blogging Behaviors. PLoS ONE 9(1), e84997 (2014), doi:10.1371/journal.pone. 0084997
5. Golbeck, J., Robles, C., Edmondson, M., Turner, K.: Predicting personality from Twitter. In: Proceedings of 2011 IEEE Third International Conference on Privacy, Security, Risk and Trust (Passat) and 2011 IEEE Third International Conference on Social Computing (Socialcom) (2011)
6. Golbeck, J., Robles, C., Turner, K.: Predicting personality with social meida. In: Proceedings of CHI 2011 Extended Abstracts on Human Factors in Computing Systems (2011)
7. Quercia, D., Kosinski, M., Stillwell, D., Crowcroft, J.: Our Twitter profiles, our selves: Predicting personality with Twitter. In: Proceedings of 2011 IEEE Third International Conference on Privacy, Security, Risk and Trust (Passat) and 2011 IEEE Third International Conference on Social Computing (Socialcom) (2011)
8. Bai, S., Hao, B., Li, A., et al.: Predicting Big Five Personality Traits of Microblog Users. In: 2013 IEEE/WIC/ACM International Joint Conferences on Web Intelligence (WI) and Intelligent Agent Technologies (IAT), vol. 1, pp. 501–508. IEEE (2013)
9. Rapoza, K.: China's Weibos vs US's Twitter: And the Winner Is? Forbes (May 17, 2011) (retrieved August 4, 2011)
10. Gosling, S.D., Augustine, A.A., Vazire, S.: Manifestations of personality in online social networks: Self-reported facebook-related behaviors and observable profile information. Cyberpsy-chology, Behavior, and Social Networking 14(9), 483–488 (2011)
11. Costa Jr., P.T., McCrae, R.R.: Revised NEO Personality Inventory (NEO-PI-R) and NEO Five-Factor Inventory (NEO-FFI) manual. Psychological Assessment Resources, Odessa (1992)
12. Tupes, E., Christal, R.: Recurrent personality factors based on trait ratings. Journal of Personality 60(2), 225–251 (1992)
13. McCrae, R., John, O.: An introduction to the five-factor model and its applications. Journal of Personality 60(2), 175–215 (1992)
14. Digman, J.: Personality structure: Emergence of the five-factor model. Annual Review of Psychology 41(1), 417–440 (1990)
15. John, O.: The Big Five factor taxonomy: Dimensions of personality in the natural language and in questionnaires. In: Handbook of Personality: Theory and Research, vol. 14, pp. 66–100 (1990)

16. McCrae, R.: Why I advocate the five-factor model: Joint factor analyses of the NEO-PI with other instruments. In: Personality Psychology: Recent Trends and Emerging Directions, pp. 237–245 (1989)
17. Schmitt, D., Allik, J., McCrae, R., Benet-Martinez, V.: The geographic distribution of Big Five personality traits: Patterns and profiles of human self-description across 56 nations. Journal of Cross-Cultural Psychology 38(2), 173 (2007)
18. Schrammel, J., Köffel, C., Tscheligi, M.: Personality traits, usage patterns and information disclosure in online communities. In: BCS HCI 2009: Proceedings of the 2009 British Computer Society Conference on Human-Computer Interaction, pp. 169–174. British Computer Society, Swinton (2009)
19. Pennebaker, J., Francis, M., Booth, R.: Linguistic inquiry and word count: LIWC 2001. Lawrence Erlbaum Associates, Mahway (2001)
20. Hall, M., Frank, E., Holmes, G., Pfahringer, B., Reutemann, P., Witten, I.: The WEKA data mining software: An update. ACM SIGKDD Explorations Newsletter 11(1), 10–18 (2009)

Identifying Opinion Leaders from Online Comments

Yi Chen, Xiaolong Wang, Buzhou Tang[*], Ruifeng Xu, Bo Yuan, Xin Xiang,
and Junzhao Bu

Key Laboratory of Network Oriented Intelligent Computation,
Harbin Institute of Technology Shenzhen Graduate School, Shenzhen 518055, China
{chenyi,xuruifeng,yuanbo,xiangxin,bujunzhao}@hitsz.edu.cn,
wangxl@insun.hit.edu.cn, tangbuzhou@gmail.com

Abstract. Online comments are ubiquitous in social media such as micro-blogs, forums and blogs. They provide opinions of reviewers that are useful for understanding social media. Identifying opinion leaders from all reviewers is one of the most important tasks to analysis online comments. Most existing methods to identify opinion leaders only consider positive opinions. Few studies investigate the effect of negative opinions on opinion leader identification. In this paper, we propose a novel method to identify opinion leaders from online comments based on both positive and negative opinions. In this method, we first construct a signed network from online comments, and then design a new model based on PageTrust, called TrustRank, to identify opinion leaders from the signed network. Experimental results on the online comments of a real forum show that the proposed method is competitive with other related state-of-the-art methods.

Keywords: Opinion Leader, Online Comments, Signed Networks, PageRank.

1 Introduction

The social media, such as micro-blogs, forums and blogs, have rapidly developed during the past decades. Online comments provide an important place for reviewers to share their positive or negative opinions toward affairs or products. They have become an important component of social media. Among reviewers, there are several opinion leaders whose opinions greatly affect others. Identifying these opinion leaders from online comments is of great significance. For governments, opinion leaders can help positive opinions toward hot events to be spread rapidly, which will promote social harmony and stability; for enterprises, with the help of opinion leaders, new products can be quickly spread to their customs and achieve a good sale; for publics, knowing opinion leaders means mastering the mainstream viewpoints about hot events or new products.

Many methods have been proposed to identify opinion leaders in social networks. The early methods simply use statistical measurements based on social network analysis, including degree centrality [1], closeness centrality [2], graph centrality [2] and

---

[*] Corresponding author.

H.-Y. Huang et al. (Eds.): SMP 2014, CCIS 489, pp. 231–239, 2014.
© Springer-Verlag Berlin Heidelberg 2014

betweenness centrality [2]. The shortcoming of these statistical measurements is that they may result in finding junk opinion leaders who forge a deluge of links as they only consider network links. Subsequently, a number of relatively complex methods, such as PageRank, HITS [3, 4], TwitterRank [5] and PageRank-like algorithm [6], are proposed for opinion leader identification.

The main limitation of these methods is that they only consider negative opinions, which is not suitable for online comments. Recently, three PageRank-like models are proposed for ranking nodes of networks with negative links, i.e., the Simple Page-Rank (Sim-PR) [7], Virtual PageRank (Vir-PR) [8] and PageTrust (PT) [9], which are potential to identify opinion leaders from online comments.

In this paper, we propose a novel method to identify opinion leaders from online comments. In this method, we first construct a user network with positive and nega-tive links (called signed network) via four procedures: setting up a basic weight post network with explicit and implicit links, labeling the sign of explicit links, inferring the sign of implicit links, transforming the signed post network into a signed user network. Then we design a new model based on PageTrust, called TrustRank, to iden-tify opinion leaders from the signed network. Compared with other methods, our me-thod considers both positive and negative opinions. In addition, the negative link has two meanings: "negation" sense and "weak-positive" sense. Negation sense means leaving and stopping. Weak-positive sense, for example "the enemy of my enemy is my friends", means keeping and going on. Experimental results on the online com-ments of a real forum show that the proposed method is competitive with other related state-of-the-art methods.

The remaindering sections of this paper are organized as follows: Section 2 describes how to construct a signed network from online comments. Section 3 intro-duces a novel model based on PageTrust to identify opinion leaders in the signed network. Section 4 discusses the experiments on an online comment dataset from a real forum. Conclusions are drawn in section 5.

2   Construct Signed Networks

Before illustrating the detail procedures, we define some basic notations. Let $P=\{p_1,p_2,...,p_n\}$ be a comment post set and p_i represents the ith post. Let $U=\{u_1,u_2,...,u_m\}$ be a user set and u_j represents the jth user. Let u_{p_i} denotes a user who posts p_i. In addition, let $G_P(P,E_P)$ represents a comment post signed network, where E_P denotes the relationship between posts (explicit or implicit). Each edge $p_{ij}\in E_P$ can be expressed by a four-tuple (p_i,p_j,w_{ij},s_{ij}), where w_{ij} denotes the edge weight ranging from (0, 1], s_{ij} denotes the edge sign ranging from $\{-1, 1\}$. Similarly, let $G_U(U, E_U)$ represents a user signed network and each edge $u_{ij}\in E_U$ can be denoted by a four-tuple (u_i,u_j,w_{ij},s_{ij}). Note that commonly the number of set P is not tanta-mount to the set U because some users may have two or more posts.

2.1   Network Construction

Explicit Link. In an online forum, the explicit link is denoted by two meta operations: reply and citation. For posts p_i and p_j, if p_j is a direct reply toward or cite p_i,
there exist an explicit link from p_j to p_i. Note that if p_m which cites p_j is a reply to p_i,
there is   an explicit link from p_m to p_j rather than p_i. And the weight value is 1 for all
explicit links.

Implicit Link. In an online forum, p_i and p_j $(1 \leq i, j \leq n)$ have not an explicit relationship, however, they share some semantic similarities. There exist an implicit link from
p_j to p_i because before posting, users commonly have read and also been influenced
by several preceding posts. We adopt a method for measuring post similarity [10] to
calculate the relevancy between two posts. The implicit link weight is equal to the
post similarity.

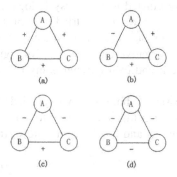

Fig. 1. Illustration of structural balance theory.
Triads with odd number of pluses are labeled as
balanced (A and B) and Triads with even positive
edges are labeled as unbalanced (C and D).

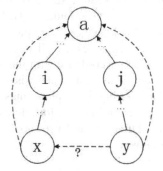

Fig. 2. A virtual triad axy. s_{yx} denotes
the implicit link sign, s_{xa} and s_{ya}
denotes virtual link sign.

2.2   Network Labeling

We classify comment posts into two types: direct and indirect posts. The direct posts
show a direct attitude toward the target post author, including "excellent post",
"agreement" or "junk post"; the indirect posts illustrate sentiment opinions toward the
role talked in the target post. It is not difficult to differentiate them because the direct
posts have several particular characteristics: laconicism; slang; accompanied by special punctuations.

   Two strategies are adopted for labeling the direct and indirect posts. For the former, we use two artificial sets collected according to the special characteristics. This
is practicable because in a certain media-sharing the direct posts are often enumera-
ble. For the indirect posts, three procedures are necessary for labeling: first setting up
a heterogeneous triad for sharing a common object, then utilizing the sentiment analysis to label the direct attitude, at last multiplying the direct sign.

Sentiment Analysis. Firstly, the comment post is split into several sentences. Secondly, six artificial features are extracted and a common classification method is selected for identifying the sentence sentiment orientation. Lastly, the post sentiment orientation is in accordance with the sign of the large number of sentence orientation. All features are set 1 or 0 according to whether they appear or not. The six artificial features include: ①8664 positive words, we collect 2036 positive words manually and use these words as seeds for expending in the Chinese thesaurus set (tongyici cilin) and the final set contains 8664 positive words; ②16894 negative words; ③13 negation words; ④169 degree words; ⑤130 degree words, such as "think"; ⑥16 special punctuations, such as "!!!!!" and "????".

2.3   Network Inferring

Structural Balance. Structural balance theory is originated in social psychology and then formulated by Heider in the 1940s [11]. Figure 1 shows several triad examples of the structural balance. A simple triad is balanced based on two kinds of situations: the sign of three edges are all positive; there are two negative and one positive signs. They are in accordance with the intuition that "the friend of my friend is my friend" and "the enemy of my enemy is my friend".

For inferring the implicit links labels, it is necessary to build up a virtual triad. As can be seen in figure 2, for inferring the label of implicit link s_{yx}, we need to build up a virtual triad axy and also should know the link sign s_{xa} and s_{ya}. We formulate s_{xa} in a general way:

$$s_{xa} = s_{x,...,i,...,a} \tag{1}$$

For calculating the sign $s_{x,...,i,...,a}$,

$$s_{x,...,i,...,a} = s_{xx+1} \cdots s_{i-1i} \cdots s_{a-1a} \tag{2}$$

Similarly, s_{ya} can be written as:

$$s_{y,...,j,...,a} = s_{yy+1} \cdots s_{j-1j} \cdots s_{a-1a} \tag{3}$$

So, s_{yx} equals to:

$$s_{yx} = s_{x,...,i,...,a} \cdot s_{y,...,j,...,a} \tag{4}$$

2.4   Network Transforming

Now we transform the post signed network into user signed network. Given an edge $p_{ij} \in E_P$ for transform, there always exist three situations: ① $u_{p_i} = u_{p_j}$; ② $u_{p_i} \neq u_{p_j}$ and $u_{ij} = 0$; ③ $u_{p_i} \neq u_{p_j}$ and $u_{ij} \neq 0$. For situation one, we just leave out the edge because u_{p_i} and u_{p_j} are the same user. For situation two, we set $u_{ij} = p_{ij}$. There is a little more complicated in situation three. If $|u_{ij}| = 1$ and $|p_{ij}| < 1$ we just leave out the edge in that "1" which expresses explicit links should be more reliable than implicit links. Otherwise we set $u_{ij} = p_{ij}$ because the later commonly represents the relationship new status.

3   Identifying Opinion Leaders

Intuitively, the opinion leaders from online comments can be interpreted similar to the "authority" of a web page: A user has high influence if the sum of influence of his/her comments is high. Here, based on the PageTrust method [12], we propose a TrustRank method for handling positive and negative weight links.

The TrustRank is grounded on an idea that the distrust(negative) links may strengthen the possibility of leaving the network. In a $n \times n$ distrust matrix P, P_{ik} denotes the proportion of walkers in node i who distrusts node k. And the diagonal of P gives the proportion of walkers that distrust the node they are in. In that manner, $(1-P_{ii})$ represents the proportion of remaining walkers in node $i \in n$. Accompanied by the distrust matrix, the iteration process is defined as:

$$x_i^{(t+1)} = (1-P_{ii}) \cdot \left[\alpha \sum_{j,(j,i) \in \ell^+} x_j^{(t)} \cdot w_{ji} / D_j + (1-\alpha)z_i \right],$$ (5)

where $\left[\alpha \sum_{j,(j,i) \in \ell^+} x_j^{(t)} / d_j + (1-\alpha)z_i \right]$ represents the traditional PageRank process and $(1-P_{ii})$ denotes the possibility of remaining the graph. The dynamic iteration of distrust matrix P is updated according to the equation:

$$P^{\widetilde{(t+1)}} = T^{(t)} \cdot P^{(t)} ,$$ (6)

where T is the transition matrix and $T_{ij}^{(t)}$ is the ratio of node i who was j at time t,

$$T_{ij}^{(t)} = \frac{\alpha A_{ji}^+ x_j^{(t)} \cdot w_{ji} / D_j + (1-\alpha)z_i x_j^{(t)}}{\alpha \sum_{k,(k,i) \in \ell^+} x_k^{(t)} \cdot w_{ki} / D_k + (1-\alpha)z_i}$$ (7)

In a signed network, there are three types of propagation information: positive-positive, negative-positive or reverse, negative-negative. The PageTrust method has considered the former two but ignores the negative-negative information. Here we introduce a new matrix to model the negative-negative propagation information. There are four types of atomic propagations in a network: direct propagation, co-citation, transpose trust and trust coupling [9]. Let M denote a connection matrix in a signed network with n nodes, the corresponding operators of four atomic propagations are: $M, M^T M, M^T, MM^T$. Let $\beta = (\beta_1, \beta_2, \beta_3, \beta_4)$ be a vector weight for the four atomic propagations. Then four atomic propagations can be combined into a single matrix $C_{M,\beta}$:

$$C_{M,\alpha} = \beta_1 M + \beta_2 M^T M + \beta_3 M^T + \beta_4 MM^T ,$$ (8)

where $\beta_1 + \beta_2 + \beta_3 + \beta_4 = 1$. The kth iterative propagation can be denoted as $C_{M,\alpha}^k$. By introducing diverse damping factors (γ_t, γ_d) for trust and distrust matrix propagation, the detail matrix iterative propagation process can be defined as:

$$C_{M,\beta(i,j)}^k = \sum_{s=1}^{n} \gamma C_{M,\beta(i,s)}^{k-1} \cdot \gamma C_{M,\beta(s,j)}$$ (9)

And the γ can be formulated as:

$$\gamma = \begin{cases} \gamma_t & C_{M,\beta(i,j)}^k > 0 \\ \gamma_d & C_{M,\beta(i,j)}^k < 0 \end{cases} \tag{10}$$

Based on formulae (9) and (10), the negative-negative propagation information can be calculated by:

$$F_{(i,j)} = \sum_{k=1}^{m} \sum_s \gamma_d C_{M,\beta(i,s)}^{k-1} \cdot \gamma_d C_{M,\beta(s,j)}, \tag{11}$$

where m denotes the iteration depth and its value depends on actual situations. Then we combine the matrix F with the original matrix M to construct a new information matrix M' which has obtained the negative-negative information:

$$P_{(i,j)}^{(0)} = \begin{cases} M_{(i,j)} & if\ M_{(i,j)} \neq 0, \\ F_{(i,j)} & if\ i \neq j, \\ 0 & if\ i = j. \end{cases} \tag{12}$$

4   Experiments

4.1   Datasets

The datasets are collected from the category of "Online Military Review" of the ChinaNet Military Forum[1] which is the largest and also the most active military forum in China. The forum provides a vote button for forum visitors to share agreements. We use the agreements for the gold opinion leaders. Here, we randomly downloaded about 1000 threads on 7 May, 2013. Then removing those comment posts which is less than two pages and get 53 threads. We extract some useful information, including user ID, post content, post floors and post votes (Since the crawler algorithm failed to download the vote information, we manually record the top 10 opinion leaders who have the most votes).

Table 1. Comparisons of top 10 opinion leaders between four models in thread 4

	UserID(votes)	Sim-PR	Vir-PR	PT	TR
1	Zjs16(2221)	1	2	8	3
2	Xysgy(1380)	6	10	11	9
3	Sfpy(867)	4	4	9	1
4	Fs_KK(562)	117	109	3	14
5	Afhdhg(277)	2	5	13	4
6	Yzqf618(173)	20	19	12	7
7	Dh_wgd(169)	21	20	14	8
8	Lsw(162)	7	6	17	5
9	Jlh(151)	48	44	7	13
10	Kw(151)	32	30	1	12

---

[1] http://club.china.com/data/threads/12171906/index.html

4.2   Results

We compare the TrustRank (TR) model with three models for handling positive and negative links, including the Sim-PR, Vir-PR and PT. In addition, to give a clear comparison of the ranking result, we adopt Mean Absolute Percentage Error(MAPE) and F-measure. The MAPE is a common method for evaluating the difference between actual values and predicting values:

$$MAPE = \frac{1}{n}\sum_{i=1}^{n}\left|\frac{A_i - F_i}{A_i}\right|,$$ (13)

The lower the MAPE value, the better the ranking. The F-measure is a well-known evaluation method in information retrieval. The higher the F-measure, the better the result. Given the precision P and recall R, the F-measure is defined as:

$$F = \frac{2PR}{P + R}$$ (14)

For illustrating the ranking result of four models, we present the top 10 opinion leaders of the thread 4 as shown in table 1. The second column is the real top 10 opinion leaders and the last four columns are the ranking order of them in four models. The table illustrates that the TR model has a better ranking than the other three models. Specifically, 7 out of 10 nodes have a better order than the PT model; 5 out of 10 nodes obviously outperform both Sim-PR and Vir-PR models, while the other 5 nodes have an imminent ranking.

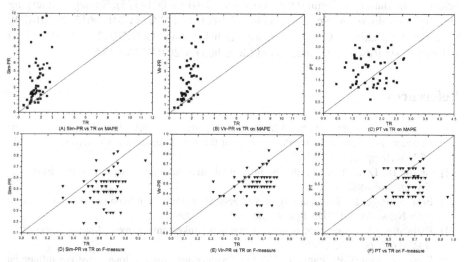

Fig. 3. Comparisons between TR and other three models in forum datasets. The top three figures show the results on MAPE; the bottom three figures show the results on F-measure.

Applying MAPE and F-measure for measuring the top 10 opinion leaders of 53 threads, figure 3 illustrates the detail results of four models. The figure shows that the TR model outperforms the other three models in two evaluation methods.

Specifically, 26 out of 53 threads' TR ranking is optimal in four models using MAPE and F-measure. 43 out of 53 threads' TR ranking is optimal in four models using MAPE or F-measure. For the remaining 10 threads, although they fail to obtain an optimal solution in TR model, they all get a suboptimal value using MAPE or F-measure. This lie in two reasons: 1) The TR model formulates negative links as negative influence, which degrades nodes that accept a large number of negative links the possibility of being important nodes. This satisfies the common sense that one opposed by a majority of people is less likely to being opinion leaders. So, the TR model outperforms both Sim-PR and Vir-PR models which treats negative links as none or positive influence. 2) The TR model also treats negative links as weak-positive influence, which can make up for mis-degrading nodes that accept negative links occasionally. So, the TR model is superior to PT model which only takes negative influence into consideration.

5   Conclusions

In this paper, we propose a novel method to identify opinion leaders from online comments based on both positive and negative opinions. The effectiveness of this method is validated on the online comments of a real forum.

Acknowledgements. This paper is supported in part by grants: NSFCs (National Natural Science Foundation of China) (61173075, 61272383 and 61370165), Natural Science Foundation of Guang Dong Province (S2013010014475), Strategic Emerging Industry Development Special Funds of Shenzhen (ZDSY20120613125401420), Shenzhen International Cooperation Research Funding (GJHZ20120613110641217) and Key Basic Research Foundation of Shenzhen (JC201005260118A).

References

[1] Zhang, J., Ackerman, M.S., Adamic, L.: Expertise networks in online communities: Structure and algorithms. In: Proceedings of the 16th International Conference on World Wide Web, pp. 221–230 (2007)

[2] Ghosh, R., Lerman, K.: Predicting influential users in online social networks. Eprint arXiv: cs/1005.4882

[3] Brin, S., Page, L.: The anatomy of a large-scale hypertextual Web search engine. Computer Networks and ISDN Systems 30, 107–117 (1998)

[4] Kleinberg, J.M.: Authoritative sources in a hyperlinked environment. Journal of the ACM (JACM) 46, 604–632 (1999)

[5] Weng, J., Lim, E.P., Jiang, J., He, Q.: TwitterRank: finding topic-sensitive influential twitterers. In: Proceedings of the Third ACM International Conference on Web Search and Data Mining, pp. 261–270 (2010)

[6] Saez-Trumper, D., Comarela, G., Almeida, V., Baeza-Yates, R., Benevenuto, F.: Finding trendsetters in information networks. In: Proceedings of the 18th ACM SIGKDD International Conference on Knowledge Discovery and Data Mining, pp. 1014–1022 (2012)

[7] Richardson, M., Agrawal, R., Domingos, P.: Trust management for the semantic web. In: Fensel, D., Sycara, K., Mylopoulos, J. (eds.) ISWC 2003. LNCS, vol. 2870, pp. 351–368. Springer, Heidelberg (2003)

[8] Tai, A., Ching, W., Cheung, W.: On Computing Prestige in a Network with Negative Relations. International Journal of Applied Mathematical Sciences 2, 56–64 (2005)

[9] Guha, R., Kumar, R., Raghavan, P., Tomkins, A.: Propagation of trust and distrust. In: Proceedings of the 13th International Conference on World Wide Web, pp. 403–412 (2004)

[10] Gabrilovich, E., Markovitch, S.: Computing semantic relatedness using wikipedia-based explicit semantic analysis. In: Proceedings of the 20th International Joint Conference on Artificial Intelligence, pp. 1606–1611 (2007)

[11] Heider, F.: Attitudes and cognitive organization. The Journal of Psychology 21, 107–112 (1946)

[12] De Kerchove, C., Dooren, P.: The PageTrust algorithm: how to rank web pages when negative links are allowed. In: SIAM: Data Mining Proceedings (2008)

A Method of User Recommendation in Social Networks Based on Trust Relationship and Topic Similarity[*]

Yufeng Ma, Zidan Yu, and Jun Ding

Department of Computer Science and Engineering
East China University of Science and Technology
Shanghai, 200237, China
045120100@mail.ecust.edu.cn

Abstract. In the research area of user recommendation in social network sites (SNS), there exist problems that some algorithms based on the structure of SNS are resulting in low quality recommendation results due to lack of model and mechanism to express users' topic similarity, some algorithms which use topic model to measure the theme similarity between users cost a lot of time because of the topic model have a high time complexity in case of large amounts of data. This paper proposed a hybrid method for user recommendation based on trust relationship and topic similarity between users, aiming to widening their circle of friends and enhancing user stickiness of SNS. Two main steps are involved in this process: (1) a trust-propagation based community detection method is proposed to model the users' social relationship; (2) a topic model is applied to retrieve users' topics from their microblogging, and gain the recommendations by the topic similarity. Our research brings two major contributions to the research community: (1) a Peer-to-Peer trust model, PGP, is introduced to the field of community detection and we improve the PGP model to compute trust value more precise; (2) a distributed implementation of the topic model is proposed to reduce total execution time. Finally, we conduct experiments with Sina-microblog datasets, which shows the model we proposed can availably compute the trust degree between users, and gain a better result of recommendation. Our evaluation demonstrates the effectiveness, efficiency, and scalability of the proposed method.

Index Terms: trust degree, community detection, topic model, topic similarity, user recommendation.

1 Introduction

As the development of Social Network Site, a very large scale of relationship graph is formed between users. In recent years, one of the most popular social network sites should be Sina microblog. At the end of 2013, Sina microblog has nearly 600 million registered users, 60 million daily active users, and 200 million daily microblogs. As for the reasean why SNS can ensure users to be active, it is not noly because they can maintain the real friend relationship, but also because they can find more attentive users and expand their social circle. Therefore, user recommendation plays an important role in SNS, its effect has a direct influence of the popularity of social networking sites.

---

[*] This work is supported by the science and technology support program (2013BAH11F03).

H.-Y. Huang et al. (Eds.): SMP 2014, CCIS 489, pp. 240–251, 2014.

Much of the recent researchs of social network sites' user recommendation mainly focuse on two fileds: one is based on users' characteristics, and the other is based on users' relationship graph. User recommendation methods which based on user characteristics can fall into three categories: content based, common topics based, user labels based. [1] used personal informations of users as the content and calculated similarity for user recommendation; Piao S [2] applied term vectors extracted from users' tweets to represent user's topics, then recommended friends with similar topics; Gou L [3] used the association rules to compute the label similarity between users for recommendation.

However, the algorithms based on users' characteristics can only recommend similar friends and the recommendation is very tedious for users. [4] built a graph by the following relationship between users and proposed a link prediction method for recommendation based on the structure similarity. [5] utilized the adjacency relationships between users to calculate the similarity matrix, then recommended friends by relevant information about the users' network topology. The method based on users' relationship graph always recommend familiar users, and it will not be unable to recommend potential friends with the same topics accurately.

At the same time, the recent research on friend recommendation did not make full use of the trust relationship between users, a closest friend with a high trust degree did not play a proper role for recommendation in these methods because closest friends were considered to have the same effect as unfamiliar friends.

PGP (Pretty Good Privacy) is a model which uses the asymmetric encryption to protect data security. Unlike the trust model those use CA (Certificate Authority) as a trusted third party, PGP leaves the trust of the initiative to users and utilizes the recommender trust model to measure the trust level between users. However, there are two obvious drawbacks when PGP is applied in large-scale social networks:

1. The trust chain can not include nodes more than two, that means it only has two types of edge in the trust chain: one of this is direct trust and the other is recommender trust. It is obviously not enough for large-scale social network

2. PGP gives only three or four trust grades, so it is difficult to distinguish the trust degree between users precisely.

For these problems, we propose a hybrid method for user recommendation based on the trust degree and topic similarity between users. Firstly, we present an improved PGP model, the trust propagation model (**TPM**), to calculate the trust-degree matrix between users, and a trust-degree based community detection method is established. Secondly, we pick out the community which has the target user, and retrieve all users' topics from their microblogging by the topic model in this community, then recommend friends to the target user by the topic similarity.   •

2   Whole Framework

The algorithm in this paper aims at recommending potential friends for the SNS users. We realize there are two kinds of connections between users in the SNS through the previous analysis, one is the link in networks between users, and the other is users' topic similarity in all kinds of topics. The goal of our algorithm is to combine these two kinds of connections. The overview of our algorithms is shown in Figure 1.

Firstly we divide users into groups by the community detection algorithm, then extract the user's topic distribution in the community and calculat the topic similarity between the target user and all other users in this community, finally give the user's personalized friends recommendation. There is a brief introduction to these steps..

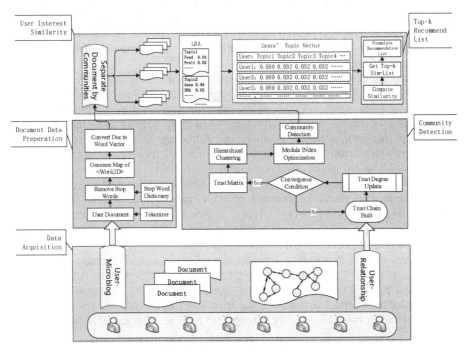

Fig. 1. Framework overview

1. Community discovery algorithm

For the existing user relationship, based on the PGP trust model we propose a modified method, to measure the trust between nodes, and then use fast hierarchical clustering algorithm which proposed by J. H. Ward [6] , with the aim to optimize module of the whole social network degree, to generate the final community. The specific calculation process is in section 3.

2. Users' topic similarity calculation

After the community division, the network topology relationship of the users in the community become more compact, then a user's microblogs can be seen as one document, This document contains the topic of the user's attention, and we adopt LDA model which proposed by Blei.D.M [7] to calculate each document topic distribution, then calculate the topic similarity of each document. The specific calculation process is in section 4.

3. Generate friend recommended list

After user topic similarity calculation for recommended users, apart from its users, select the Top-K greatest similarity users, to generate recommendation list. At this point, we have completed the whole process of personalized friends' recommendation.

3   The Community Detection Algorithm Based on User Trust Chain

Considering the actual situation of the social network, people give more trust to their close friends, and a group of users, in which people have high trust degree between each other, will form a community. Referencing the presenter trust models in the PGP, this paper propose a modified community discovery algorithm based on trust-chain, which can overcome the two drawbacks of PGP: trust chain maximum length is two, and trust measure size is not enough.

3.1   The Definition of Community in SNS

To build a social network un-weighted digraph $G = (V, E)$, V denotes vertex (user) collection, $|V|=n$, E denotes the collection of user relationship, e_{ij} denotes the edge linked v_i, v_j. Such a social network graph can also be denoted as the adjacency matrix $A=a_{ij}(v_i, v_j \in V)$, when $(v_i, v_j) \in E$, $a_{ij}=1$.

Divided G into k, get a partition which has k vertex collections, $\varphi = \{N_1, N_2... N_k\}$. If $N_i \in \varphi$, all the nodes' trust chain in N_i are intensive, and all the nodes' trust chain out of N_i are sparse, then φ is the community partition of G which based trust chain.

3.2   The Calculation of Trust in Trust Chain

If there is an edge between v_i and v_j, it donates user i has followed user j, it also donates i has a trust with j. If the total trust of each user is 1, and he equally assigns it to all the users he has followed. Initialize the trust:

If i has followed j, and i has followed $\sum_{k=1}^{n} a_{ik}$ users, then the initialized trust of i to j is:

$$Tru(i, j) = \frac{1}{\sum_{k=1}^{n} a_{ik}} \tag{1}$$

Assuming that trust can be spread in two ways: series and parallel:

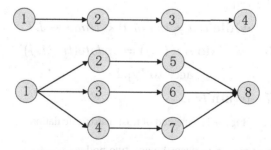

Fig. 2. The propagation of trust degree

For the series of trust propagation, the trust of each path will decrease with the increase of the intermediate node,

$$Tru(1,4) = Tru(1,2) \times Tru(2,3) \times Tru(3,4)$$

For parallel transmission of trust, the trust will increase with the increase of the path,

$$Tru(1,8) = sum \begin{cases} Tru(1,2) \times Tru(2,5) \times Tru(5,8) \\ Tru(1,3) \times Tru(3,6) \times Tru(6,8) \\ Tru(1,4) \times Tru(4,7) \times Tru(7,8) \end{cases}$$

Combined with the two assumptions, given the following method to update the trust between i,j:

1. Given the depth of the threshold value, through the breadth-first search, get m trust chains from i to j.

2. For these m trust chains, the intermediate nodes of each trust chain is $v` = \{v1`, v2`...vk`\}$, $v1`$ denotes i, $vk`$ denotes j. Then the collection of trust TChain = $\{Tru(v1`,v2`), Tru(v2`,v3`)...Tru(vk-1`,vk`)$, $TChain_q^p$ denotes the q trust in trust chain p.

3. Add all of the m trust chains, then the trust between i and j is:

$$Tru(i, j) = \sum_{p=1}^{m} \prod_{q=1}^{k} TChain_q^p \qquad (2)$$

Then give the following iteration method, to calculate trust between any two nodes:

GET THE TRUST-MATRIX

1   **for** $(i,j) \in \{(v_i, v_j) | v_i, v_j \in V, v_i \neq v_j\}$

2   **do** initialize $tru(i,j)$ acc. to Eq.1

3   times $\leftarrow 1, LT \leftarrow limit\ Times$

4   $tru' = 0$

5   **while** $tru \neq tru'$ or $0 < times < LT$

6   **do** $tru'(i,j) = update(tru(i,j))$
         acc. to Eq.2

7   return tru

Fig. 3. The method of trust matrix calculation

So we get the trust matrix to describe any two nodes.

Similarly, we consider the trust between node i and community C, it defined as the average trust of node i and all the nodes in community C:

$$Tru(i,C) = \frac{1}{|C|}\sum_{j \in C} Tru(v_i, v_j) \tag{3}$$

|C| denotes the nodes' number of community C. The trust between community and community, defined as trust root mean square average of any two nodes in the community :

$$Tru(C_1, C_2) = \sqrt{\frac{\sum\limits_{i \in C_1}\sum\limits_{j \in C_2} Tru^2(v_i, v_j)}{|C_1| + |C_2|}} \tag{4}$$

3.3   The Steps of Hierarchical Clustering

In the last section, this paper puts forward how to calculate the trust between two nodes. The next question is how to cluster these nodes to form the community. Here, we use the bottom-up hierarchical clustering algorithm which proposed by J. H. Ward [6], the specific algorithm is as follows:

Divide all the nodes of G , get community partition $\varphi = \{\{v\}, v \in V\}$, each community has only one node. Calculate all adjacent nodes trust first, and then through the following step iteration to combine the community:

HIERARCHICAL CLUSTERING FOR COMMUNITY
1   $TRU = \{tru(C_i, C_j) | C_i, C_j \in \phi, i \neq j\}$
2   **while** $size(\phi) \neq 1$
3   **do** Choose $\{(C_1, C_2) | tru(C_1, C_2) = \max(TRU)\}$
4          $C_3 \leftarrow C_1 \cup C_2$
5          $\phi \leftarrow (\phi - \{C_1, C_2\}) \cup C_3$
6          $TRU \leftarrow \{tru(C_i, C_j) | C_i, C_j \in \phi, i \neq j\}$

Fig. 4. The hierarchical clustering procedures of community

After the process above, to get the final community partition $\varphi_n = \{V\}$. This is a process of hierarchical clustering, according to the order of nodes are merged into the community, dendrogram can be constructed. The leaves of the dendrogram is all the vertices on the social network graph, and the internal nodes of the dendrogram corresponds to the "merge" steps of the algorithm: just that corresponds to a new community which merges its two children nodes.

3.4   Module Index

Module index Q [8] is used to depict community features of strength. Its main idea is based on "network community structure, the more obvious, the greater the difference between it and the random network". In general, a higher degree of module of network represent the partitioning effect is better. For the above given hierarchy clustering figure, consider some form of division, graph will be divided into k communities. m denotes the total number of network connections, m_i denotes the number of network connections in C_i, d_i denotes the sum of the node degrees in community i. The definition of Q function is the difference between the actual number of connections and expectation connections in randomly connected network:

$$Q = \sum_{i=1}^{k} \left[\frac{m_i}{m} - (\frac{d_i}{2m})^2 \right] \tag{5}$$

If Q tends to be 1, it donates the community has the very strong structure. In the actual network, the value is generally between 0.3 ~ 0.7. Finally using the hierarchical clustering results, according to the maximum module degrees of cutting, then get the community partition.

4   Extract User Topic Distribution and Similarity Calculation

On the microblogging platform, the main performance way of users interaction is the microblog contents. Microblog contents host the user's will, goals, and even social relationship. How to use topic learning and unsupervised clustering, according to these microblog contents, get the user's hobby then clustering the users who have similar microblog contents is a core problem in social network data mining field. In topic models, the most common is LDA (Latent Dirichlet Allocation) model.

Because each microblog is short, in order to extract users' topics accurately and rapidly, this paper will treat each user's all microblogs as a document. Then we use LDA model to extract the topics of these documents, to get each user corresponding topic distribution, ultimately through the cosine similarity calculation, getting the topic of the similarity between the users, the greater the value shows that the better correlation between the two users.

4.1   Latent Dirichlet Allocation Topic Model

LDA (Latent Dirichlet Allocation) is a three layers of Bayesian probability model. The basic idea is that documents are represented as random mixtures over latent topics, where each topic is characterized by a distribution over words. It assumes there are D document and K topics in the corpus. For each document d in D, there are N_d words. These K topics are shared by all the documents, but for each document has a corresponding topic distribution. For each topic has a corresponding word distribution. LDA assumes the following generative progress for each document d in a corpus D:

CREAT A DOCUMENT

1   **for** each topic $k \in \{1...k\}$
2   **do** draw $\varphi \sim Dir(\beta)$
3   **for** each document $d \in \{1...d\}$
4   **do** draw $\theta_d \sim Dir(\alpha)$
5   **for** each word $w \in \{1...N_d\}$
6   **do** draw $\xi_{d,n} \sim Mult(\theta_d)$
7   draw $W_{d,n} \sim Mult(\psi_{\xi_{d,n}})$

Fig. 5. The generative process of a document

Figure 6 shows the probabilistic graphical modeled LDA:

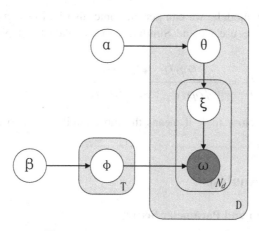

Fig. 6. The probabilistic graphical model of LDA

As the figure makes clear, there are three level to the LDA representation. The parameters α and β are corpus-level parameters, assumed to be sampled once in the process of generating a corpus. The variables θ_d are document-level variables, sampled once per document. Finally, the variables ζ and ω are word-level variables and are sampled once for each word in each document.

The topic distribution θ_d of document d is sampled from Dirichlet distribution. The word distribution φ_ζ of ζ topic is sampled from Dirichlet distribution. Both of the Dirichlet distributions are independent. θ for example, its corresponding distribution function is as follows:

$$p(\theta \mid \alpha) = \frac{\Gamma(\sum_{i=1}^{T} \alpha_i)}{\prod_{i=1}^{T} \Gamma(\alpha_i)} \prod_{t=1}^{T} \theta_t^{\alpha_t - 1} \tag{6}$$

T denotes the number of topics, α_i ($i \in [1,T]$) is each component of α, Γ is the gamma function.

We obtain the probability of a corpus:

$$p(D \mid \alpha, \beta) = \prod_{d=1}^{M} \int p(\theta_d \mid \alpha) p(w_{dn} \mid z_{dn}, \beta)) d\theta_d \tag{7}$$

Now require the distributions of φ and θ. Here we use Gibbs extraction, it is easy to implement and can effectively extract topic from the large-scale corpus. It's the most popular LDA model extraction algorithm. Gibbs extraction iterates sampling the topic of words in each position instead of directly computing φζ and θd. Once the topic of words in each position is definite, φ ζ and θd can be calculated. Then we can get the topic distribution of documents, and the word distribution of topics.

4.2   Topic Similarity Calculation

Assuming that from LDA model, we can get the topic distribution $Topic_A$, $Topic_B$ of user A and B. Then we use Cosine Similarity to calculate the topic similarity between A and B.

$$sim(A, B) = \frac{Topic_A \bullet Topic_B}{\mid Topic_A \mid \times \mid Topic_B \mid} \tag{8}$$

If Cosine similarity is close to 1, it means the more similar the two topic of the user's attention.

5   Experimental Analysis

5.1   Experimental Data and Parameter Settings

Tang Jie, the professor in Tsinghua University, has crawled 1.7 million users and 0.4 billion following relationships among them [9]. In this paper, we use following method to extract instances: 1. five adjacent user nodes are selected as the seeds; 2. In each iteration, the depth priority method is used to fetch the nodes which far away from the current user; the breadth first method is used to fetch the adjacent nodes of current users. Finally, we obtain a social network with 1538 nodes, and 732643 relationships.

This paper consider the maximum length of trust chain is 4 due to the complexity of the algorithm. When calculate the distribution of the users documents, we use ICTCLAS as the tokenizer, and get 65023 words totally. LDA is conditioned on three parameters. In this paper, they are set as T=30, $\alpha = 50/T$ and $\beta = 0.01$ according to [10].

5.2   Evaluation Method

In the experiment, we recommend *Top-K* friends for users in the offline data set by the algorithm posted in this paper. In order to verify the test results, this paper adopts the following recommendation system evaluation standards: *precision* rate, *recall* rate. Let $R(u)$ denotes the list of users that are recommended to u, and $T(u)$ denotes the list of friends of u. The *precision* and *recall* are defined as follows:

$$precision = \frac{\sum_{u \in U} \mid R(u) \bigcap T(u) \mid}{\sum_{u \in U} \mid R(u) \mid} \tag{9}$$

$$recall = \frac{\sum_{u \in U} \mid R(u) \bigcap T(u) \mid}{\sum_{u \in U} \mid T(u) \mid} \tag{10}$$

F-measure is the harmonic mean of *precision* and *recall:*

$$F = \frac{2 \times precesion * recall}{precesion + recall} \tag{11}$$

5.3   Experimental Results and Analysis

Choose simple based on link relations and simple based on the topic method as a benchmark experiment. In order to get the final result, the algorithm proposed in this paper will compare with two kinds of benchmark experiments. Table 1 show the results, R denotes *Recall* rate, P denotes *Precision* rate, *F* denotes harmonic average.

According the results of table 1, we can get several evaluation index charts of the algorithm, as shown in figure[7-9]:

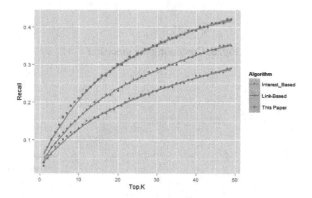

Fig. 7. Comparing the recall

Table 1. The comparison result of different algorithm

Top-K	This Paper			Link-Based			Topic-Based		
	R	P	F	R	P	F	R	P	F
5	9. 01%	25. 43%	0. 090	6. 22%	14. 82%	0. 062	4. 12%	7. 86%	0. 041
10	10. 46%	20. 67%	0. 105	6. 65%	11. 25%	0. 067	4. 18%	6. 12%	0. 042
15	10. 52%	17. 75%	0. 105	6. 31%	9. 15%	0. 063	3. 91%	5. 13%	0. 039
20	10. 32%	15. 78%	0. 103	5. 91%	7. 92%	0. 059	3. 64%	4. 51%	0. 036
25	9. 96%	14. 33%	0. 100	5. 49%	6. 94%	0. 055	3. 40%	4. 05%	0. 034
30	9. 61%	13. 22%	0. 096	5. 12%	6. 24%	0. 051	3. 19%	3. 70%	0. 032
35	9. 26%	12. 31%	0. 093	4. 78%	5. 68%	0. 048	3. 00%	3. 42%	0. 030
40	8. 97%	11. 62%	0. 090	4. 50%	5. 22%	0. 045	2. 85%	3. 19%	0. 028
45	8. 64%	10. 96%	0. 086	4. 22%	4. 82%	0. 042	2. 69%	2. 98%	0. 027
50	8. 42%	10. 53%	0. 084	4. 03%	4. 55%	0. 040	2. 59%	2. 85%	0. 026

As the figures shown, both precision rate and recall rate of this algorithm is significantly better than the benchmark   algorithms, The reason is that in this paper, we consider both the links between users and topic information of the user's interests, which is better than to consider only a single aspect of the informations.

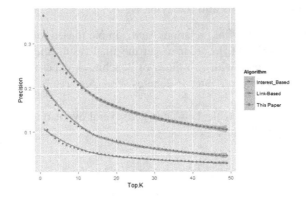

Fig. 8. Comparing the precision

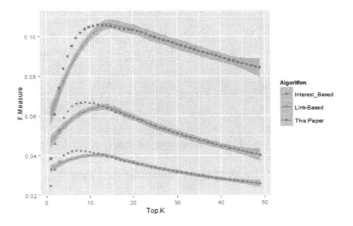

Fig. 9. Comparing the F value

6   Conclusion

This paper presents a novel method to recommend users in social networks, in which both link-based similarity and topic-based similarity are considered. The method exploits a two-phase process to provide the trust based and topic sensitive recommendations for social network users. Firstly, we utilize the trust recommend models in PGP to model the trust relations between the user nodes, compute the trust degree between each pair of users, divided the whole social network into several sub-community networks based on the trust degree. Secondly, we use LDA to model the topics of each users in a community and analyze the topic similarity between users, finally, we recommend the potential friends by the rank of these similarity. Experimental results on real-world data demonstrate that the proposed method outperforms other algorithms in terms of precision, recall, and F-measure. In our future work, we plan to deploy our method onto the real social network site. This will allow us to collect users' feedbacks on our results, which can be helpful for us to adjust the recommendation strategies.

References

[1] Jeckmans, A., Tang, Q., Hartel, P.: Privacy-Preserving Profile Similarity Computation in Online Social Networks. In: Proceedings of the 18th ACM Conference on Computer and Communications Security, pp. 793–796. ACM Press, New York (2011)

[2] Piao, S., Whittle, J.: A feasibility study on extracting twitter users' topics using NLP tools for serendipitous connections. In: 2011 IEEE Third International Conference on Privacy, Security, Risk and Trust (passat) and 2011 IEEE Third International Conference on Social Computing (socialcom), pp. 910–915. IEEE (2011)

[3] Gou, L., You, F., Guo, J., et al.: SFViz: topic-based friends exploration and recommendation in social networks. In: Proceedings of the 2011 Visual Information Communication-International Symposium, p. 15. ACM (2011)

[4] Yin, D., Hong, L., Xiong, X., et al.: Link formation analysis in microblogs. In: The 34th International ACM SIGIR Conference on Research and Development in Information Retrieval (SIGIR 2011), Beijing, China, pp. 1235–1236 (July 2011)

[5] Armentano, M., Godoy, D., Amandi, A.: Recommending Information Sources to Information Seekers in Twitter. In: Proceedings of the IJCAI: International Workshop on Social Web Mining, Barcelona, Spain (2011)

[6] Ward Jr., J.H.: Hierarchical grouping to optimize an objective function. Journal of the American Statistical Association 58(301), 236–244 (1963)

[7] Blei, D.M., Ng, A.Y., Jordan, M.I.: Latent dirichlet allocation. The Journal of Machine Learning Research 3, 993–1022 (2003)

[8] Newman, M.E.J., Girvan, M.: Finding and evaluating community structure in networks. Physical Review E 69(2), 026113 (2004)

[9] Zhang, J., Liu, B., Tang, J., et al.: Social influence locality for modeling retweeting behaviors. In: IJCAI 2013 (2013)

[10] Weng, J., Lim, E.P., Jiang, J., et al.: Twitterr rank: finding topic sensitive influential twetterers. In: Proc. of the 3rd ACM International Conference on Web Search and Data Mining (2010)

Author Index